Tapio Salmi, Johan Wärnå, José Rafael Hernández Carucci, César A. de Araújo Filho
Chemical Reaction Engineering

Also of Interest

Advanced Process Engineering Control.
Agachi, Cristea, Csavdari, Szilagyi, 2016
ISBN 978-3-11-030662-0, e-ISBN 978-3-11-038816-9

Process Synthesis and Process Intensification.
Methodological Approaches
Rong (Ed.), 2017
ISBN 978-3-11-046505-1, e-ISBN 978-3-11-046506-8

Product and Process Design.
Driving Innovation
Harmsen, de Haan, Swinkels, 2018
ISBN 978-3-11-046772-7, e-ISBN 978-3-11-046774-1

Process Intensification.
Design Methodologies
Gómez-Castro, Segovia-Hernández (Eds.), 2019
ISBN 978-3-11-059607-6, e-ISBN 978-3-11-059612-0

Product-Driven Process Design.
From Molecule to Enterprise
Zondervan, Almeida-Rivera, Carmada, 2020
ISBN 978-3-11-057011-3, e-ISBN 978-3-11-057013-7

Tapio Salmi, Johan Wärnå,
José Rafael Hernández Carucci,
César A. de Araújo Filho

Chemical Reaction Engineering

A Computer-Aided Approach

DE GRUYTER

Authors
Prof. Dr Tapio Salmi, Prof. Dr Johan Wärnå, Dr José Rafael Hernández Carucci,
Dr César A. de Araújo Filho
Åbo Akademi University
Laboratory of Industrial Chemistry and Reaction Engineering
Biskopsgatan 8
FI-20500 Turku/Åbo
Finland

ISBN 978-3-11-061145-8
e-ISBN (PDF) 978-3-11-061160-1
e-ISBN (EPUB) 978-3-11-061250-9

Library of Congress Control Number: 2019955792

Bibliographic information published by the Deutsche Nationalbibliothek
The Deutsche Nationalbibliothek lists this publication in the Deutsche Nationalbibliografie;
detailed bibliographic data are available on the Internet at http://dnb.dnb.de.

© 2020 Walter de Gruyter GmbH, Berlin/Boston
Cover image: Suriyapong Thongsawang / Moment / Getty Images
Typesetting: Integra Software Services Pvt. Ltd.
Printing and binding: CPI books GmbH, Leck

www.degruyter.com

Preface

Chemical reactors form the core of any chemical process: 100% of industrially applied chemical processes take place in chemical reactors. The products of chemical processes are numerous: fuels, fertilizers, lubricants, plastics, glass and ceramics, components for electronics, detergents, perfumes, drugs, bioproducts.... The list is endless. Modern production technology must be sustainable and clean, which implies that selective routes for chemical synthesis are preferred. The chemical production equipment, the reactor, should operate in a predictable and optimized way.

The big issue in predicting the behaviour of existing reactor systems and designing new ones is mathematical modelling based on mass, energy and momentum balances. However, models alone can produce imaginary reactors only. The models need their parameters, such as rate and equilibrium constants, mass and heat transfer coefficients etc. Models appearing in connection with chemical reactors are typically highly non-linear in their nature. This fact emphasizes the role of numerical computing: analytical solutions to balance equations are limited to simple – mostly isothermal – cases. Analytical solutions to classical, mainly linear problems in chemical reaction engineering are treated in detail in many excellent textbooks of the field.

The aim of this book is not to compete with existing textbooks in chemical reaction engineering, but to illustrate how models of chemical reactors are built up in a systematic manner, step by step, how the numerical solution algorithms for the models are selected and finally, how computer codes are written for numerical performance. Nowadays, the programming language selected for use in reactor modelling is almost a matter of taste. Among many options, we have stuck with some of them: Matlab, Python and Fortran. Examples solved with Matlab and Python demonstrate how various reactor problems can easily be simulated with a high-level programming language. This approach has the advantage that the engineer can focus all his efforts on the physical and chemical problem itself. Most problems can be solved by merging the appropriate program components together. To solve advanced reactor problems, a fundamental programming language, such as Fortran or C++, is still needed.

Chemical engineers working with chemical reactors in the industry and in academia recognize the current reality: chemical reaction engineering is a rapidly developing area; more and more complex systems are involved in new sophisticated reactors such as monoliths, reactors with structural packings, multiphase fluidized beds and microreactors. Simultaneously, sophisticated modelling concepts are applied to well-established reactor systems, such as tank reactors and fixed beds. The existing commercial process simulation software can help us to some extent, but particularly in developing new process concepts and improving the present ones, tailored reactor modelling and numerical simulation are inevitable steps. Thus, a professional of chemical reaction engineering should know the basic principles of reactor modelling, including kinetics, thermodynamics, transport phenomena and flow modelling and the solution of models by numerical techniques. In this book, we provide the physical–chemical fundamentals of reactor modelling and mathematical models for some standard designs of chemical reactors. In our opinion, this basic knowledge and way of thinking is needed for the modelling of new, avant-garde reactors.

Many excellent books have been published to describe existing chemical processes along with reactor and separation equipment. For this reason, the qualitative description of various chemical reactors is kept to a minimum, at a level necessary for understanding the basis of mathematical modelling. Similarly, the multitude of correlation equations appearing in chemical reaction engineering and transport theory is not reviewed in detail. Instead, some simple, classical correlations are presented in order to demonstrate how correlation equations are integrated in the modelling of chemical reactors.

Recently, highly exact flow modelling achieved by solving Navier-Stokes equations has become fashionable. The concept called computational flow dynamics (CFD) provides a detailed idea of the fluid velocity field in the vessel. Coupling flow modelling with chemical reactions is a very demanding task and thus worthy of a separate treatment. Therefore, the CFD approach to reactor modelling is not discussed in this book. Furthermore, the authors believe that a major share of chemical reactor modelling will also follow the classical pathway in future: description of reaction kinetics and thermodynamics, mass and heat transfer effects coupled with models for a priori determined flow patterns.

We hope that this book will attract readers both from academia and the industry, for example, postgraduate students and researchers who wish to initiate a reactor modelling project of their own as well as engineers dealing with reactor modelling and design in their everyday work.

This interdisciplinary book is the result of long-term fruitful collaboration among chemical engineers, numerical mathematicians and computer experts. The material has been gradually developed and presented on international postgraduate courses in chemical reaction engineering at Åbo Akademi University.

The authors are grateful to numerous members of the team at the Laboratory of Industrial Chemistry and Reaction Engineering, Åbo Akademi, who have contributed to the concept of computer-aided chemical reaction engineering: Dr Jari Romanainen, Dr Sami Toppinen, Dr Juha Lehtonen, Dr Esko Tirronen, Dr J.-P. Mikkola, Dr Mats Rönnholm, Dr Fredrik Sandelin, Dr Sebastien Leveneur, Dr Matias Kangas, Dr Henrik Grénman, Dr Teuvo Kilpiö and Dr Vincenzo Russo. Without their enthusiasm and devotion and the experience and scientific results they contributed, this book would never have seen the light of day. Mr Daniel Wärnå made a valuable contribution to the final word processing of the text. Last but not least, Professor Heikki Haario from Lappeenranta University of Technology has made an enormous contribution to the development of parameter estimation procedures for chemical engineering – he is behind the implementation of many exercises presented in this book. We are grateful to our families for steady support during this long task.

In Kuressaare/Saaremaa, Estonia, at Bishop's castle, March 2016,
The Authors

Contents

Preface —— V

Nomenclature —— XV

1 Introduction —— 1

2 Kinetics in reaction engineering —— 5
2.1 Stoichiometry of multiple reactions —— 5
2.2 Reaction kinetics in chemical reaction engineering —— 6
2.2.1 General concepts —— 6
2.2.2 Examples of rate equations —— 7

3 Modelling of homogeneous systems —— 9
3.1 Mass balances for completely backmixed tank reactors – batch, semi-batch and continuous operation —— 9
3.2 Mass balances for tubular reactors —— 14
3.3 Energy balances of homogeneous systems —— 20
3.3.1 Tank reactor —— 23
3.3.2 Tubular plug flow reactor —— 25
3.3.3 Batch reactor —— 26
3.3.4 Semi-batch reactors —— 27
3.4 Physical properties and correlations of homogeneous systems —— 28
3.4.1 Heat capacity and reaction enthalpy —— 28
3.4.2 Pressure drop in tubular reactors —— 29
3.4.3 Dispersion coefficient —— 30
3.5 Numerical solution of homogeneous reactor models —— 31
3.5.1 Model structures and algorithms —— 31
3.5.2 Software build-up —— 36

4 Modelling of fixed beds and fluidized beds —— 41
4.1 Simultaneous reaction and diffusion in fluid films and porous media —— 42
4.2 Catalytic fixed bed reactors —— 45
4.2.1 Models for fixed beds —— 46
4.2.2 Pseudo-homogeneous models for fixed beds —— 47
4.2.3 Heterogeneous model for fixed beds —— 52
4.2.3.1 Special case —— 58
4.2.4 Model equations for the bulk phase —— 60
4.2.5 Pressure drop in fixed beds —— 64

4.3	Numerical solution of fixed bed models —— 64	
4.3.1	Solution of pseudo-homogeneous models —— 64	
4.3.2	Solution strategy of heterogeneous models —— 66	
4.4	Catalytic fluidized beds —— 68	
4.4.1	Modelling approaches to fluidized beds —— 68	
4.4.2	Kunii-Levenspiel model of fluidized beds —— 71	
4.4.2.1	Bubble phase —— 71	
4.4.2.2	Cloud and wake phases —— 72	
4.4.2.3	Emulsion phase —— 72	
4.4.2.4	Energy balances —— 73	
4.4.2.5	Parameters in Kunii-Levenspiel model —— 73	
4.5	Numerical solution of fluidized bed models —— 74	
4.6	Physical properties and correlations for catalytic two-phase systems —— 76	
4.6.1	Effective diffusion coefficients in a gas phase —— 76	
4.6.2	Mass and heat transfer coefficients around solid particles —— 77	
4.6.3	Mass transfer coefficients for fluidized beds —— 78	

5   Modelling of three-phase systems —— 81

5.1	Mass balances of three-phase reactors —— 82
5.1.1	Phase boundaries —— 82
5.1.2	Liquid-phase mass balances —— 84
5.1.3	Gas-phase mass balances —— 87
5.1.4	Tank reactors with complete backmixing —— 88
5.1.5	Catalyst particles in three-phase reactors —— 89
5.1.6	Slurry reactor in the absence of mass transfer resistances —— 91
5.2	Energy balances of three-phase reactors —— 92
5.3	Numerical aspects —— 93

6   Modelling of gas–liquid systems —— 97

6.1	Gas–liquid contact —— 99
6.2	Gas and liquid films —— 101
6.2.1	Mass balances for films —— 101
6.2.2	Energy balances for liquid films —— 105
6.3	Gas–liquid tank reactors —— 107
6.4	Gas–liquid column reactors —— 108
6.4.1	Boundary conditions for balance equations —— 112
6.5	Energy balances for gas–liquid reactors —— 112
6.6	Physical properties of gas–liquid systems —— 113
6.6.1	Diffusion coefficients in gas and liquid —— 114
6.6.1.1	Gas phase —— 114
6.6.1.2	Liquid phase —— 114

| 6.6.2 | Gas–liquid equilibrium —— **117** |
| 6.7 | Numerical strategies for gas–liquid reactor models —— **118** |

7 **Equipment and models for laboratory experiments —— 123**
7.1   Homogeneous batch reactor —— **123**
7.2   Homogeneous stirred tank reactor (CSTR) —— **127**
7.3   Catalytic fixed bed in integral mode —— **128**
7.4   Catalytic differential reactor —— **129**
7.5   Catalytic gradientless reactor —— **130**
7.6   Catalytic slurry reactor —— **131**
7.7   Classification of laboratory reactor models —— **131**
7.7.1   Algebraic and differential models —— **132**
7.7.2   Linearity and non-linearity of the model —— **132**

8 **Parameter estimation in reaction engineering —— 135**
8.1   Principles of non-linear regression analysis —— **135**
8.2   Statistical and sensitivity analysis of parameters —— **139**
8.3   Suppression of correlation between parameters —— **142**
8.3.1   Correlation in rate expressions —— **143**
8.3.2   Correlation in temperature dependencies —— **145**
8.4   Systematic deviations and normalization of experimental data —— **147**
8.5   Direct integral method —— **152**
8.6   Parameter estimation from non-isothermal data —— **157**
8.7   Estimation of parameters from semi-batch experiments —— **159**
8.7.1   Composite reactions in the presence of a heterogeneous catalyst —— **160**
8.7.2   Composite reactions in the presence of a homogeneous catalyst —— **164**

Bibliography —— 169

Exercises

E.1   **Gas-phase tube reactor —— 175**

E.2   **Synthesis of maleic acid monoester in a semi-batch reactor —— 176**

E.3   **Exothermic reaction in a continuous stirred tank reactor —— 177**

E.4   **Production of phtalic anhydride in a fixed bed reactor —— 178**

E.5   Water–gas shift in a fixed bed reactor – diffusional limitations —— 180

E.6   Steady-state CSTR's in series: oxidation of Iron(II) to Iron(III) —— 182

E.7   A fluidized bed reactor —— 184

E.8   Three-phase slurry reactor: Hydrogenation of aromatics —— 185

E.9   Chlorination of p-cresol in a continuous stirred tank reactor —— 187

E.10 Reaction between methanol and triphenyl methyl chloride —— 188

E.11 Use of millireactor for the kinetic study of very fast reaction: Dehydrochlorination of 1,3-dichloro-2-propanol —— 189

E.12 Multiple liquid-phase reaction system —— 191

E.13 Gas–liquid reactions in a semi-batch reactor —— 195

E.14 Gas-phase reaction in a differential reactor —— 197

E.15 Three-phase reactions in a semi-batch reactor —— 200

E.16 Non-isothermal liquid phase reaction in a CSTR —— 203

E.17 Oxidation of sulphur dioxide in an optimal multi-bed reactor system —— 205

E.18 Modelling of a monolith channel —— 206

E.19 Heterogeneous two-dimensional model for a catalytic fixed-bed reactor —— 207

E.20 Dissolution of a solid particle in a batch reactor —— 208

Appendices

A.1   Numerical strategies in the solution of non-linear algebraic equations and ordinary differential equations —— 211
A.1.1   Non-linear algebraic equations —— 211
A.1.2   Ordinary differential equations —— 212

A.1.2.1   Semi-implicit Runge-Kutta methods —— 213
A.1.2.2   Linear multistep methods —— 215
           References —— 217

A.2   Computer simulation of CSTR, PFR and batch reactor models —— 218
Example 1 —— **218**
Example 2 —— **221**

A.3   Numerical simulation of non-isothermal tubular reactors —— 225

Index —— 231

Nomenclature

a	shape factor
a	area-to-volume ratio
A	frequency factor
A	area, particularly area accessible to mass and heat transfer
$A, B, \ldots$	general coefficients
c	concentration
c_p	mass-based heat capacity
c_{mP}	mole-based heat capacity
d	diameter
D	diffusion or dispersion coefficient
D_e	effective diffusion coefficient
E	density function
E_a	activation energy
f	function
F	frequency function
ΔH	reaction enthalpy
g	function in regression analysis
ΔG	Gibbs free energy
G	mass flow per cross-section
h	heat transfer coefficient
j	factor for heat and mass transfer
k	rate constant
k_F, k_L, k_G	mass transfer coefficients
K	equilibrium constant
K	mass transfer coefficient in Kunii-Levenspiel model
L	reactor length
m	mass
M	molar mass
M	heat flux
n	amount of substance
n'	flow of amount of substance
N	diffusion flux
p	parameter
P	pressure
Q	objective function in regression
r	generation rate
r	radial coordinate
R	reaction rate
R	gas constant 8.3143 J/(mol·K)
R_c	particle radius, particle characteristic dimension
s	shape factor for a particle; $s = a - 1$
S	surface
t	time
T	temperature
V	volume
V_m	molar volume
V'	volumetric flow rate

w	fluid velocity
w	weight factor
x	mole fraction
x	independent variable
y	dependent variable
z	length coordinate
α, β	general coefficients and exponents
γ	activity coefficient
δ	thickness
ε	porosity
η	effectiveness factor
θ	dimensionless time
λ	heat conductivity
μ	dynamic viscosity
ν_{ij}	stoichiometric coefficient for component i in reaction j
ν	kinematic viscosity
ξ	dimensionless radial coordinate
ξ	extent of reaction
ρ	density
τ	space time, residence time
τ	tortuosity, labyrinth factor
Φ	association factor
ω	weight factor in direct integral method

Dimensionless numbers

Bo	Bodenstein number
Da	Damköhler number
Ha	Hatta number
Nu	Nusselt number
Pe	Péclet number
Pr	Prandtl number
Re	Reynolds number
Sc	Schmidt number
Sh	Sherwood number
φ, Φ	Thiele modulus

Subscripts and superscripts

A, B . . .	general index for components
b	bulk
bc	bubble cloud
B	bed or bulk (ρ_B = catalyst bulk density)
BET	Brunauer-Emmett-Teller equation
ce	cloud emulsion
eq	equilibrium
exp	experimental

G	gas
i	component index
j	reaction index
L	liquid
m	molar property
mf	minimum fluidisation
p	constant pressure
P	particle
R	radial or reactor
s	surface
T	tube
v	constant volume
*	active site on catalyst surface

1 Introduction

A chemical reactor is a piece of equipment in which the chemical transformation of raw material, the reactants, into products takes place. The chemical reactor is only one part of the entire chemical process, and typically it is preceded in the process by separation units where the raw material is purified. Similarly, the chemical reactor is seldom able to produce pure products from the raw material, as some raw materials remain unreacted and by-products are formed. Therefore, the desired products are separated from the undesired ones in a separation unit after the chemical reactor, as illustrated in the principal flow sheet presented in Figure 1.1.

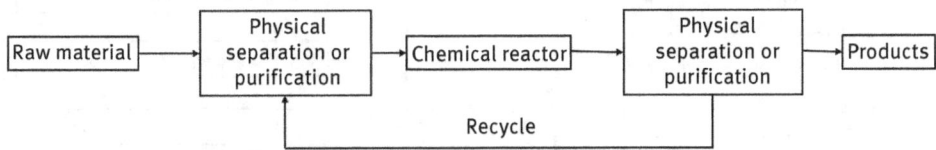

Figure 1.1: Principal flowsheet of a chemical process.

The unreacted reactant molecules are recirculated back into the chemical reactor to maintain a good economy in the utilization of raw material. Thus, the process stages described earlier represent key steps in any chemical process: preparation of the raw material for the chemical transformation, the chemical transformation itself and the separation of valuable reaction products from unreacted raw material and by-products. Several sophisticated separation techniques are at our disposal. The dominating one is distillation, but extraction, absorption, adsorption and crystallization are also commonly used. Development, description and modelling of separation processes are large subfields of chemical engineering, but they are not included in this book.

Our discussion shows that the performance of a chemical reactor is of crucial importance: the better – that is more efficiently and selectively – the chemical reactor is operated, the higher the output of the overall process. A major part of the total cost of a chemical process is usually caused by the separation operations; therefore, the general goal is to perform the reaction as well as possible in the chemical reactors to minimize the costs of separation operations and to achieve a global optimization of the process.

This book deals with the modelling and simulation of chemical reactors, which is at the core of any industrial process designed for chemical transformations. The basis of understanding it lies in the physical–chemical fundamentals, that is, reaction stoichiometry, kinetics and thermodynamics (Chapter 2). The book covers homogeneous reactors where only one phase – gas or liquid – is present (Chapter 3), heterogeneously catalysed systems (catalytic two- and three-phase reactors, Chapters 4 and 5) as well as gas–liquid systems (Chapter 6). Finally, the chemical engineer cannot limit himself to well-known systems, where all the kinetic, thermodynamic and transport parameters are known a priori. In most cases, experiments in laboratory as well as bench and pilot scales are needed. Therefore, the last chapters (7 and 8) are devoted to experiments on laboratory scale and to estimation of parameters from experimental data. The crucial steps of modelling and simulation are illustrated in Figure 1.2.

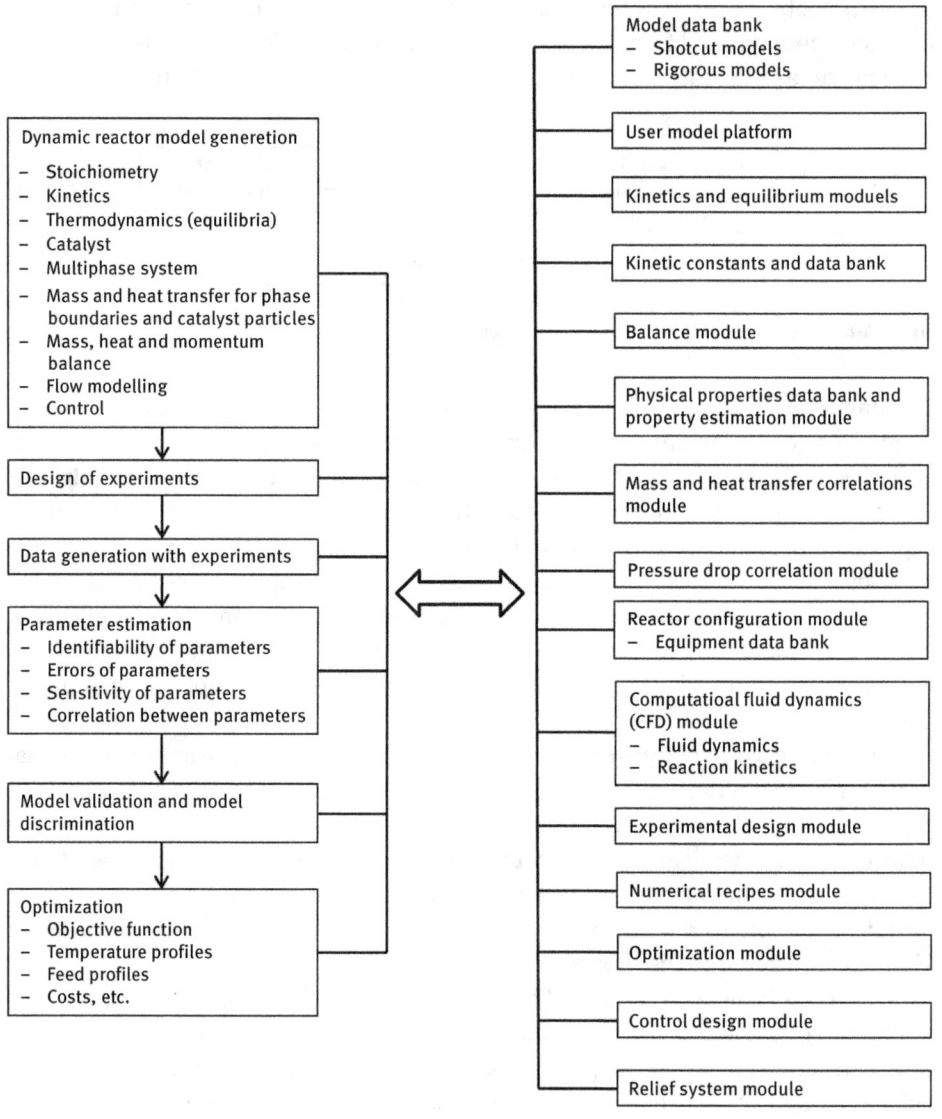

Figure 1.2: Steps of modelling and simulation work (after Tirronen and Salmi (2003)).

Efficient and robust numerical algorithms and computer software play a central role in the successful solution of reactor models. The enormous development of computing speed during the last decades has made the dreams of yesterday a reality: what theoreticians scribbled on paper in the past can be computed exactly today. Real numerical computing of processes enables a more precise, economical and environmentally sustainable process design. We try to provide the reader with a systematic approach and a guided journey from reactor modelling through numerical methods and computer implementation to the final calculation of the reactor performance. Have a nice trip!

2 Kinetics in reaction engineering

Modelling of chemical processes in general and chemical reactors in particular is based on stoichiometry, thermodynamics and kinetics. Reaction stoichiometry tells us the relative amounts of molecules interacting in a chemical process, the quantities of reactants needed for a chemical reaction and the number of product molecules formed. Thermodynamics determines the ultimate conversion limit of a reactant in a chemical reaction; de facto thermodynamic calculations provide the equilibrium composition of a reacting mixture. Chemical kinetics qualitatively and quantitatively explains the velocities, rates of chemical processes, that is, how fast the equilibrium composition is approached in a real chemical reactor.

2.1 Stoichiometry of multiple reactions

The basic concepts of reaction stoichiometry and reaction kinetics are briefly described in this chapter. The definition of reaction stoichiometry is based on the following concepts. The number of components is denoted by N and the number of independent chemical reactions by S. Consequently, the stoichiometry of each reaction (j) is obtained in the equation

$$\sum_{i}^{N} v_{ij} a_i = 0, \qquad j = 1, \ldots, S \tag{2.1}$$

where v_{ij} is the stoichiometric coefficient of component i in reaction j and a_i is the chemical symbol. All of this can be written with arrays in the following very concise manner:

$$v^T a = 0 \tag{2.2}$$

In order to clarify the use of the equations presented above, an example will be considered.

Example: The industrial synthesis of methanol is carried out over a solid catalyst. The reaction of importance is

$$-CO - 2H_2 + CH_3OH = 0$$

which takes place as the main reaction. Besides the main reaction, the following side reaction occurs, since the feed to the reactor usually contains some CO_2:

$$-CO_2 - H_2 + CO + H_2O = 0$$

The vector for chemical symbols can now be fixed as

$$a = [CO \ H_2 \ CH_3OH \ CO_2 \ H_2O]^T$$

resulting in the stoichiometric matrix

$$v^T = \begin{bmatrix} -1 & -2 & +1 & 0 & 0 \\ +1 & -1 & 0 & -1 & +1 \end{bmatrix}$$

Equation (2.2) gives

$$v^T a = 0$$

which can be confirmed by looking at this particular example:

$$\begin{bmatrix} -1 & -2 & +1 & 0 & 0 \\ +1 & -1 & 0 & -1 & +1 \end{bmatrix} \begin{bmatrix} CO \\ H_2 \\ CH_3OH \\ CO_2 \\ H_2O \end{bmatrix} = 0$$

The stoichiometric matrix is highly useful in formulating the mass balances for chemical reactors, as will be demonstrated in the subsequent sections.

2.2 Reaction kinetics in chemical reaction engineering

2.2.1 General concepts

Basically, the mathematical laws of reaction kinetics arise from molecular reaction mechanisms. We will not focus on the derivation of rate equations starting from reaction mechanisms here, since this subject is treated in detail in numerous textbooks devoted to chemical kinetics (e.g. Laidler 1987, Murzin and Salmi 2016). The reaction engineering consequences of the rate equations will be discussed below, and we will present a systematic formulation of rate laws and generation rates. The concept can be easily implemented in computer codes and linked to the mass balances of chemical reactors.

The rate of reaction step j is related to the generation rate of an individual component (r_i) by

$$r_i = \sum_{ja^1}^{S} v_{ij} R_j \tag{2.3}$$

where v_{ij} denotes an element of the stoichiometric matrix. In matrix formulation, eq. (2.3) implies that

$$\underline{r} = v \underline{R} \tag{2.4}$$

Chemical kinetics gives the expressions for R. Some features are important to remember in the use of the rate expressions. These are listed below:
- When the system is in equilibrium, $R = 0$ for all reactions.
- For elementary reactions, R is obtained directly from the reaction stoichiometry; $K_c = k_+/k_-$, that is, the equilibrium constant is related to the forward and backward rate constants of the reaction.
- Many reactions are non-elementary; R can be obtained in an exact form, provided that the reaction mechanism is known, otherwise an empirical expression has to be used.
- For typical kinetic laws, R can be created automatically by the computer. An example of using a computer to obtain R is given below.

Example: Automatic computation of reaction rates in case of elementary reactions:

Input:
$$A + 2B \rightleftharpoons C$$
$$C + D \rightleftharpoons 2F$$

Output: $\underline{a}^T = [A\ B\ C\ D\ F]$

Stoichiometric matrix: $\nu = \begin{bmatrix} -1 & 0 \\ -2 & 0 \\ 1 & -1 \\ 0 & -1 \\ 0 & 2 \end{bmatrix}$

2.2.2 Examples of rate equations

Rate equations for chemical processes can be obtained by considering molecular reaction mechanisms, or in an empirical way, by measuring the reaction rates and fitting an empirical model to the data. The first approach is preferable, since rate equations based on real mechanisms allow an extrapolation beyond the experimental domain. Some general equations for reaction kinetics are useful, and they are collected in Table 2.1. As shown in this table, some general tendencies of the rate expressions are visible, for instance, the appearance of the law of mass action in the nominator of the rate expressions. The rate expressions in Table 2.1 are given with concentrations (c). For gas-phase reactions, partial pressures (p) are frequently used. For non-ideal liquid mixtures, activities are applied. The activity of a component (a_i) is obtained from the concentration or the mole fraction (x_i) by using the activity coefficient (γ_i): $a_i = \gamma_i c_i$. The theory of activity coefficients is extensive and is beyond the scope of this book. The reader is referred to literature dedicated to this topic (Reid et al. 1988). Whatever approach is used for reaction kinetics, it is important that the rate equation is in harmony with the thermodynamics, so that the rate equation always gives the correct equilibrium composition as R is set to zero in the rate expression.

2 Kinetics in reaction engineering

Table 2.1: Typical rate expressions used in chemical engineering kinetics.

Reaction kinetics	Kinetic equation
Elementary kinetics $\|v_A\|A + \|v_B\|B + \cdots \leftrightarrow \|v_P\|P + \|v_R\|R + \cdots$	$R = k\left(c_A^a c_B^b \cdots - c_P^p c_R^r \cdots / K\right)$ $a, b, \ldots = \|v_A\|, \|v_B\| \ldots$ for reactants $p, r, \ldots = \|v_P\|, \|v_R\| \ldots$ for products $k = k_+, \quad K = k_+/k_-$
Exponent law $\|v_A\|A + \|v_B\|B + \cdots \leftrightarrow \|v_P\|P + \|v_R\|R + \cdots$	$R = k' c_A^\alpha c_B^\beta c_P^\gamma c_R^\delta$ $\alpha, \beta, \gamma, \ldots$ = empirical exponents often $\alpha \neq \|v_A\|, \beta \neq \|v_B\|, \ldots$
Langmuir–Hinshelwood kinetics (heterogeneous catalytic reactions)	$R = \dfrac{k\left(c_A^a c_B^b \cdots - c_P^p c_R^r \cdots / K\right)}{\left(1 + \sum K_i c_i^{s_i}\right)^\gamma}$
$\|v_A\|A + \|v_B\|B + \cdots \leftrightarrow \|v_P\|P + \|v_R\|R + \cdots$	K_i = adsorption parameter for component i $S_i = 1$ for non-dissociative adsorption $S_i = 1/2$ for dissociative adsorption Note: Many other types of Langmuir–Hinshelwood expressions exist
Polymerization kinetics	$R = k_P \sqrt{\dfrac{f_{kd}}{k_T}} c_A^a c_B^b$ k_P = rate constant for chain propagation k_T = rate constant for termination f_{kd} = constant for chain initiation The rate expression for the polymerization is valid for chain polymerization
Enzyme kinetics	$R = \dfrac{k_2 k_{-1}(c_S - c_P/K) c_0}{k_{-1} K_M (1 + c_S/K_M) + k_2 c_P/K}$
$S + E \xrightarrow{k_1} X \xrightarrow{k_2} E + P$ S = substrate E = enzyme P = product X = active complex	$K = \dfrac{k_1}{k_{-1}} \cdot \dfrac{k_2}{k_{-2}}, \quad K_M = \dfrac{k_2 + k_{-1}}{k_1}$ c_0 = total enzyme concentration K_M = Michaelis–Menten constant

3 Modelling of homogeneous systems

The characteristic feature of homogeneous systems is the presence of a single phase, gas or liquid. The chemical process itself can be enhanced by a soluble component, a homogeneous catalyst, or it can progress spontaneously by itself or without any catalyst. Typical homogeneous catalysts are acids, bases, metal complexes and enzymes. A classic example is the esterification of a carboxylic acid with an alcohol. This reaction can be described as $RCOOH + R'OH = RCOOR' + H_2O$. The non-catalytic reaction is slow, but it can be considerably accelerated by addition of a soluble catalyst, such as a mineral acid, for instance sulphuric acid or hydrochloric acid. In any case, the system is a homogeneous one, consisting of a single liquid phase only. Another example can be taken from gas-phase processes, such as thermal cracking of hydrocarbons. As the hydrocarbon mixture is heated, the chemical bonds start to break and smaller molecules are obtained. Furthermore, a number of radical reactions take place in atmosphere and in the combustion of fuels.

Homogeneous reactors are used in batch, continuous and semi-continuous (semi-batch) modes in the industry, both for gas- and liquid-phase processes. The reaction time can vary from milliseconds to hours, depending on the chemical system. The production scale and spectrum can vary from fine chemicals and pharmaceuticals to fuel components. In any case, the same basic theory can be applied to the modelling of homogeneous reactors. Furthermore, the theory of homogeneous reactors is the basis for understanding the behaviour of heterogeneous reactors, too. Mathematical models of homogeneous reactors are solved with similar algorithms to those applied to heterogeneous systems. The discussion shows that an extensive treatment of homogeneous processes is justified, even though they are in a minority among chemical reactions applied industrially.

3.1 Mass balances for completely backmixed tank reactors – batch, semi-batch and continuous operation

In this section, the mass and energy balances of homogeneous reactors are considered. The characteristic feature for a homogeneous reactor is that one phase only – gas or liquid – is present. This implies that interfacial mass transfer effects are absent, and the mass transfer effects which might be of importance are caused by incomplete mixing. Two main categories of homogeneous reactors are treated in detail here: tank reactors and tube reactors. Typical reactor configurations are illustrated in Figure 3.1. For tank reactors, we assume complete backmixing inside the reactor vessel, whereas plug flow, laminar flow and axial dispersion models are considered for tubular reactors. The continuous, backmixed reactor is often called a continuous stirred tank reactor (CSTR). Complete backmixing is easy to achieve on laboratory scale, since a

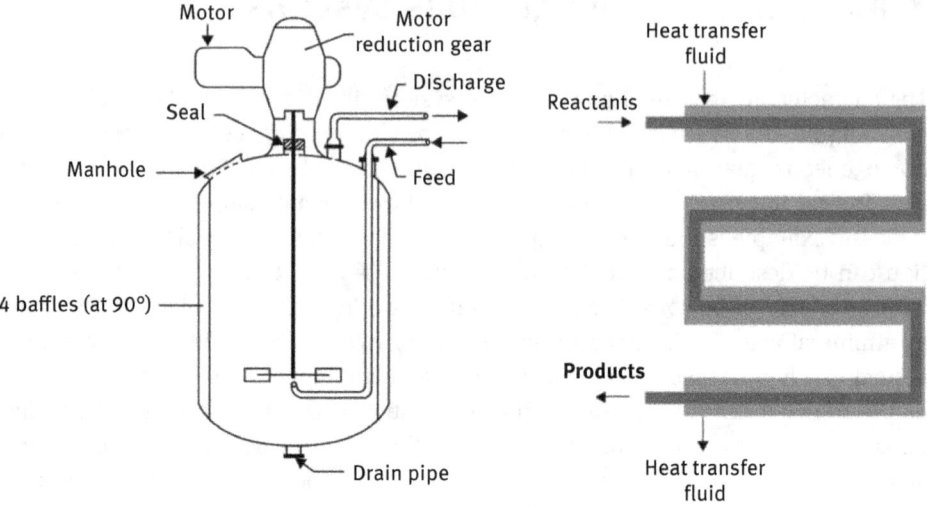

Figure 3.1: Typical tank reactor and tubular reactor configuration (after Trambouze et al. (1988)).

small vessel can be subjected to a major mixing effect (typically, the size is from few mL to 1 L), but it is much more difficult to achieve perfect mixing on production scale: the use of energy input through mixing is limited by process economics.

The general mass balance of an arbitrary component i in a tank reactor with complete backmixing is based on the following principle (Figure 3.2):

incoming 'i' + generated 'i' = outgoing 'i' + accumulated 'i'

Quantitatively, the mass balance becomes

$$\dot{n}_{0i} + r_i V = \dot{n}_i + \frac{dn_i}{dt} \tag{3.1}$$

where $\dot{n}_{0i}$ and $\dot{n}_i$ denote the inlet and outlet molar flows (Figure 3.2) and dn_i/dt describes the accumulation of component "i". The balance eq. (3.1) can be rearranged to

$$\frac{dn_i}{dt} = \dot{n}_{0i} - \dot{n}_i + r_i V \tag{3.2}$$

that is, we obtain a system of ordinary differential equations (ODEs).

For a batch reactor, the inlet and outlet flows are zero and balance eq. (3.2) is reduced to

$$\frac{dn_i}{dt} = r_i V \tag{3.3}$$

The initial condition for ordinary differential eqs. (3.2) and (3.3) is

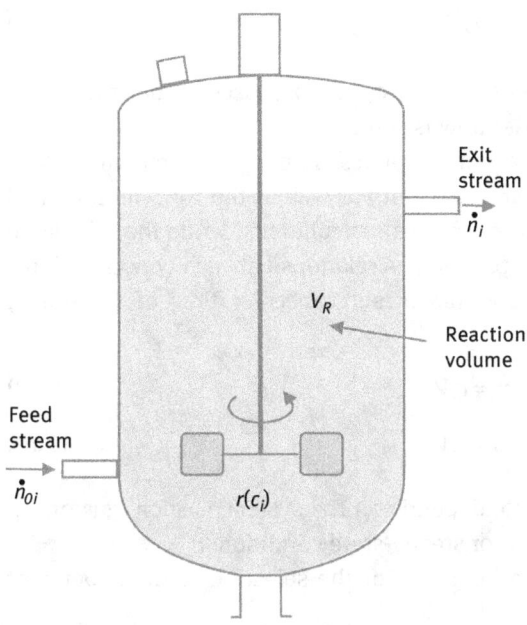

Figure 3.2: Modelling principle of a tank reactor with complete backmixing.

$$n_i = n_{0i}, \quad t = 0 \tag{3.4}$$

Below, we will consider different special cases of the general balance eq. (3.2).

A CSTR frequently operates in a steady state, which implies that $dn_i/dt = 0$ in eq. (3.2). The balance equation becomes

$$\dot{n}_i - \dot{n}_{0i} = r_i V \tag{3.5}$$

Equation (3.5) is written for each reactive component, which implies that we get a system of linear or non-linear equations (NLE). In most cases, the system is non-linear, since the generation rate terms (r_i) are typically non-linear, for example, for high-order kinetics (second- and third-order kinetics as well as complex kinetics as shown in Table 2.1).

In reactor simulations, it is often useful to express the balance equations with arrays. Equations (3.6) and (3.7) present the balances for CSTR, semi-batch reactors (SBR) and batch reactors (BR) with the aid of arrays:

$$\frac{d\underline{n}}{dt} = \underline{\dot{n}}_0 - \underline{\dot{n}} + \underline{r}V, \quad CSTR, SBR \tag{3.6}$$

$$\frac{d\underline{n}}{dt} = \underline{r}V, \quad BR \tag{3.7}$$

where the generation rate is obtained from (Chapter 2)

$$\underline{r} = \underline{\underline{v}}\,\underline{R}(\underline{c}) \tag{3.8}$$

For a SBR, either the outlet or the inlet flow is zero; in most cases, some of the reactants are fed into the SBR and the outlet flow is zero.

The balance equations contain several variables, namely the amount of substance (n_i), the flow of the amount of substance ($\dot{n}_i$) as well as the concentration (c_i). Only one of these variables is needed in practical calculations, while the others are obtained from fundamental relationships, such as relations between concentrations, volumes/volumetric flow rates and amounts of substance (or flow of amount of substance):

$$n_i = c_i V \tag{3.9}$$

$$\dot{n}_i = c_i \dot{V} \tag{3.10}$$

The selection of the dependent variable depends on the system as such (gas or liquid phase) as well as its state (transient or steady-state conditions.)

The following rules of thumb can be given for the selection of the dependent variable:
- The use of the molar flow ($\dot{n}_i$) is straightforward for all kinds of steady-state models both in gas- and liquid-phase systems.
- The use of the molar amount (n_i) is good for dynamic models, because it makes it possible to keep the ordinary differential equation system (3.2) explicit.
- Concentrations (c_i) are typically used for liquid-phase reactions, in which the change of density, that is, the reaction volume, is often negligible.

Let us consider an isothermal CSTR. The change in density and volume is ignored, that is, $\rho = \rho_0$ and $V = V_0$. This enables us to approximate the derivative $dn_i/dt = dc_i/dt \cdot V$ which is inserted in eq. (3.6). The balance equation becomes

$$\frac{d\underline{c}}{dt} = \frac{\underline{c}_0 - \underline{c}}{\tau} + \underline{r} \tag{3.11}$$

where $\tau = V/\dot{V}$ has the time unit and is called space time. For a system with a constant density, the space time coincides with the average residence time ($\bar{t}$) of the fluid. In a steady state, the time derivative becomes zero in eq. (3.11).

The characteristic features for a batch reactor are a constant reaction volume: gas-phase batch reactors are closed autoclaves and for a liquid-phase system, the change of density is often minimal. Thus, the simplified relation $dn_i/dt = dc_i/dt \cdot V$ is applied to the balance eq. (3.7), which is reduced to

$$\frac{d\underline{c}}{dt} = \underline{r} \tag{3.12}$$

Semi-batch reactors are used for liquid-phase processes, particularly for manufacturing fine and speciality chemicals. Because one or more of the reacting components are typically fed into the reactor, the volume change has to be included in the model. Provided that the density change is negligible, a simple updating expression for the liquid volume can be used:

$$V = V_0 + \int_0^t \dot{V} dt \qquad (3.13)$$

Differentiation of the amount of substance in this case gives

$$\frac{d\underline{n}}{dt} = \frac{d\underline{c}}{dt} V + \underline{c}\frac{dV}{dt} \qquad (3.14)$$

After inserting this relationship in the original balance equation, and some rearrangement we obtain

$$\frac{d\underline{c}}{dt} = \frac{\underline{c}_0 - \underline{c}}{\tau} + \underline{r}, \qquad \tau = \frac{V}{\dot{V}} \qquad (3.15)$$

which is formally equal to the balance equation of a CSTR. It should be noted, however, that the reactor volume (V) included in the space time (τ) should be updated according to eq. (3.13).

For a gas-phase CSTR operated in a steady state, the natural choice for the dependent variable is the molar flow. From eq. (3.6), we can derive the special case

$$\dot{\underline{n}}_0 - \dot{\underline{n}} + \underline{r}V = 0 \qquad (3.16)$$

Concentrations needed for the calculations of reaction rates are obtained from the trivial relations

$$\underline{c}_0 = \frac{\dot{\underline{n}}_0}{\dot{V}_0}, \underline{c} = \frac{\dot{\underline{n}}}{\dot{V}} \qquad (3.17)$$

The volumetric flow rate is updated from an equation of state, for example, $PV = ZnRT$, where the compressibility factor (Z) is a function of temperature, pressure and gas composition (Reid et al. 1988).

For a gas-phase CSTR operating under dynamic conditions, simplifying assumptions cannot be made, and we are left with the original form of the balance eq. (3.6). It is, however, necessary to express two of the three variables ($\underline{n}, \dot{\underline{n}}, \underline{c}$) with the aid of a single variable. The concentration is in this case the natural selection. Provided that the gas volume (V) is constant, which is valid in most cases, the accumulation term becomes

$$\frac{d\underline{n}}{dt} = \frac{d(\underline{c} \cdot V)}{dt} = V\frac{d\underline{c}}{dt} \qquad (3.18)$$

and we can rewrite eq. (3.6) as

$$\frac{dc}{dt} = \frac{c_0}{\tau_0} - \frac{c}{\tau} + r \tag{3.19}$$

Addition of the balances of all – even non-reactive – components gives the change of the total concentration (c) with respect to time

$$\frac{dc}{dt} = \frac{c_0}{\tau_0} - \frac{c}{\tau} + \sum r_i, \quad c = \sum c_i, \quad c_0 = \sum c_{0i} \tag{3.20}$$

A dimensionless quantity $\tau_0/\tau = \dot{V}/\dot{V}_0 = \delta$ is introduced,

$$\delta = \frac{c_0}{c} + \frac{\tau_0}{c}\left(\sum_i r_i - \frac{dc}{dt}\right) \tag{3.21}$$

De facto, the derivative of the total concentration (dc/dt) is coupled to the energy balance, since for an ideal gas ($Z = 1$) we can write

$$c = \frac{P}{RT} \tag{3.22}$$

Several simplifications of the system are possible. Let us assume that the ideal gas flow can be applied with a reasonable accuracy. For isothermal reactors operated under a constant pressure, $dc/dt = 0$ according to eq. (3.22). Consequently, $dc/dt = 0$ always in a steady state. Thus, the update of the volumetric flow rate is easily obtained for these cases,

$$\delta = \frac{c_0}{c} + \frac{\tau_0}{c}\sum_i r_i \tag{3.23}$$

If the reactor is isothermal and operates in a steady state, $c = c_0$ and the update becomes even simpler,

$$\delta = 1 + \frac{\tau_0}{c}\sum_i r_i \tag{3.24}$$

3.2 Mass balances for tubular reactors

A tubular reactor with plug flow and axial dispersion is schematically illustrated in Figure 3.3. In order to maintain the general nature of the treatment, a completely dynamic model is developed here. According to a dynamic axial dispersion model (DADM), the mass balance of an arbitrary component (i) in an infinitesimally small volume element is written as

3.2 Mass balances for tubular reactors

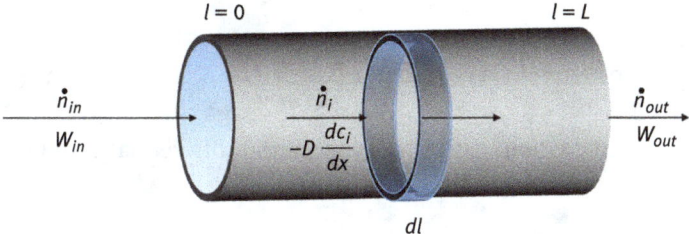

Figure 3.3: Tubular reactor with plug flow and axial dispersion.

$$\dot{n}_{i,in} + \left(-D\frac{dc_i}{dl}A\right)_{in} + r_i \Delta V = \dot{n}_{i,out} + \left(-D\frac{dc_i}{dl}A\right)_{out} + \frac{dn_i}{dt}, \quad \text{DADM} \quad (3.25)$$

where D denotes the axial dispersion coefficient. It should be noted that axial dispersion is described by a law which is formally similar to Fick's equation for molecular diffusion in dilute solutions. However, the numerical value of the dispersion coefficient might differ greatly from that of the molecular diffusion coefficients, since the axial dispersion coefficient is de facto a merged parameter combining the effects of diffusion and turbulence: turbulent eddies cause a backmixing effect in the reactor vessel. Furthermore, a global value of the axial dispersion coefficient is typically used in the axial dispersion model; not individual coefficients for different compounds. The numerical value of the axial dispersion coefficient can vary from zero (plug flow) to infinity (complete backmixing, CSTR). Correlation equations for the calculation of axial dispersion coefficients will be discussed in Section 3.4.3.

By denoting the differences as

$$\Delta \dot{n}_i = \dot{n}_{i,out} - \dot{n}_{i,in}, \quad \Delta\left(D\frac{dc_i}{dl}A\right) = \left(-D\frac{dc_i}{dl}A\right)_{out} - \left(-D\frac{dc_i}{dl}A\right)_{in} \quad (3.26)$$

we obtain

$$\Delta\left(D\frac{dc_i}{dl}A\right) + r_i \Delta V = \Delta \dot{n}_i + \frac{dn_i}{dt} \quad (3.27)$$

The accumulation term can be rewritten, since the volume of the volume-element (ΔV) remains constant:

$$\frac{dn_i}{dt} = \frac{d(c_i \Delta V)}{dt} = \Delta V \frac{dc_i}{dt} \quad (3.28)$$

The volume element is expressed by the reactor cross-section and the length coordinate:

$$\Delta V = A \Delta l \quad (3.29)$$

A rearrangement gives

$$D\frac{\Delta(dc_i/dl)}{\Delta l} + r_i = \frac{1}{A}\frac{\Delta \dot{n}_i}{\Delta l} + \frac{dc_i}{dt} \quad (3.30)$$

By letting the volume element to diminish ($\Delta l \to 0$), a partial differential equation (PDE) is obtained,

$$\frac{dc_i}{dt} = -\frac{1}{A}\frac{d\dot{n}_i}{dl} + D\frac{d^2 c_i}{dl^2} + r_i, \quad \text{ADM} \quad (3.31)$$

where the molar flow is given by

$$\dot{n}_i = c_i \dot{V} \quad (3.32)$$

The general axial dispersion model, eq. (3.31), is a parabolic . In case of negligible axial dispersion, the plug flow is obtained as a special case ($D = 0$),

$$\frac{dc_i}{dt} = -\frac{d\dot{n}_i}{dV} + r_i, \quad \text{PFR} \quad (3.33)$$

In steady-state conditions, the time derivative of the concentration vanishes, and we obtain

$$\frac{d\dot{n}_i}{dV} = D\frac{d^2 c_i}{dl^2} + r_i, \quad \text{ADM} \quad (3.34)$$

for the axial dispersion model, and

$$\frac{d\dot{n}_i}{dV} = r_i, \quad \text{PFR} \quad (3.35)$$

for the plug flow model.

For liquid-phase reactions, the density of the reaction mixture usually remains virtually constant $(\rho = \rho_0)$, which also implies that $\dot{V} = \dot{V}_0$. For such cases, the derivative of molar flow can be simplified to

$$\frac{d\dot{n}_i}{dl} = \frac{d(c_i \dot{V}_0)}{dl} = \dot{V}_0 \frac{dc_i}{dl} = w_0 A \frac{dc_i}{dl} \quad (3.36)$$

where w_0 is the superficial velocity, that is, the velocity calculated with respect to the entire cross-section of the reactor tube.

The mass balance for the axial dispersion and plug flow models thus becomes

$$\frac{dc_i}{dt} = -w_0 \frac{dc_i}{dl} + D\frac{d^2 c_i}{dl^2} + r_i, \quad \text{ADM} \quad (3.37)$$

where $D = 0$ for plug flow.

The complete axial dispersion model is a boundary value problem (BVP), the solution of which requires knowledge of boundary conditions. A classical way of obtaining the boundary conditions is to assume that axial dispersion is initiated at the reactor inlet, where plug flow is instantaneously transformed to plug flow and axial dispersion,

$$\dot{n}_{0i} = \dot{n}_i + \left(-D\frac{dc_i}{dl}A\right) \quad (3.38)$$

At the reactor inlet, the change of the volumetric flow rate is negligible, and the molar flow can be expressed as $\dot{n}_{0i} = c_{0i}\dot{V}_0$ and $n_i = c_i\dot{V}_0$. The boundary condition can be rewritten as

$$c_{0i} = c_i - \frac{D}{w}\frac{dc_i}{dl}, \quad l = 0 \quad (3.39)$$

At the reactor outlet, the boundary condition of Danckwerts (Danckwerts P. V., 1953) is normally used:

$$\frac{dc_i}{dl} = 0, \quad l = L \quad (3.40)$$

This boundary condition is thermodynamically consistent and it de facto states that axial dispersion declines at the reactor outlet.

The drawback of Danckwerts' closed boundary condition is that it somewhat artificially forces the concentration gradients to vanish at the reactor outlet, which can cause problems in the numerical solution in cases where conversion is low (i.e. considerable concentration gradients of reactants prevail) and particularly in cases where the reaction rate is increased with increasing conversion (autocatalytic kinetics) (Romanainen and Salmi 1992). Salmi and Romanainen (1995) have proposed an alternative boundary condition for the axial dispersion model, a boundary condition which interpolates between plug flow and complete backmixing as the axial dispersion coefficient increases. Sometimes the problem is circumvented by using open boundary conditions, that is, the boundary conditions are valid at infinity or require that the second derivative of the concentration – instead of the first derivative – is zero at the reactor outlet.

In connection with the axial dispersion model, the degree of axial dispersion is characterized by a dimensionless quantity called the Péclet number (referred to as the Bodenstein number in German-speaking countries), $Pe = wL/D$. The quantity appears spontaneously in the model, for example, in eq. (3.37).

For steady-state gas-phase systems, in which the change in volumetric flow rate is accounted for, the general balance eq. (3.34) is rewritten as

$$\frac{d(c_i w)}{dl} = D\frac{d^2 c_i}{l^2} + r_i \quad (3.41)$$

The derivative on the left-hand side of eq. (3.41) cannot be simplified, since the superficial velocity (w) varies because of changes in the total molar flow and/or temperature. It is thus better to approximate the dispersion term by

$$D\frac{d^2 c_i}{dl^2} = \frac{D}{w}\frac{d^2(c_i w)}{dl^2} \tag{3.42}$$

The model can now be rewritten in an explicit form with respect to the product $c_i w$, or – alternatively – the molar flow:

$$\frac{d\dot{n}_i}{dl} = \frac{D}{w}\frac{d^2 \dot{n}_i}{dl^2} + A r_i \tag{3.43}$$

With respect to the volume coordinate, we obtain

$$\frac{d\dot{n}_i}{dV} = \frac{DA}{w}\frac{d^2 \dot{n}_i}{dV^2} + r_i \tag{3.44}$$

The concept of the Péclet number is best illustrated by the steady-state axial dispersion model for the liquid phase, eq. (3.34). Provided that the superficial velocity does not change considerably inside the reactor, the mass balance eq. (3.41) can be rewritten as

$$\frac{dc_i}{dl} = \frac{D}{w}\frac{d^2 c_i}{dl^2} + \frac{1}{w} r_i \tag{3.45}$$

After introducing a dimensionless length coordinate, $= l/L$, $dl = L dz$, we get

$$\frac{dc_i}{dz} = \frac{D}{wL}\frac{d^2 c_i}{dz^2} + \tau_0 r_i \tag{3.46}$$

where $\tau_0 = L/w$. The dimensionless parameter $D/(wL)$ is called the degree of dispersion and its reciprocal value is the Péclet number for axial dispersion,

$$Pe = \frac{wL}{D} \tag{3.47}$$

The Péclet number can be estimated based on reactor characteristics, which is very helpful in evaluating the role of axial dispersion (Section 3.4.3).

The steady-state axial dispersion model is typically used for calculating the reactor performance in the presence of chemical reactions, while the non-steady-state axial dispersion model is used in the absence of chemical reactions, in order to characterise the residence time distributions of tubular reactors. In the absence of chemical reactions ($r_i = 0$), the partial differential eq. (3.37) is linear with respect to the concentration and an analytical solution can be obtained for pulsewise and stepwise introduction of an inert trace component. These are highly valuable techniques for the characterization of residence time distributions (Figure 3.4).

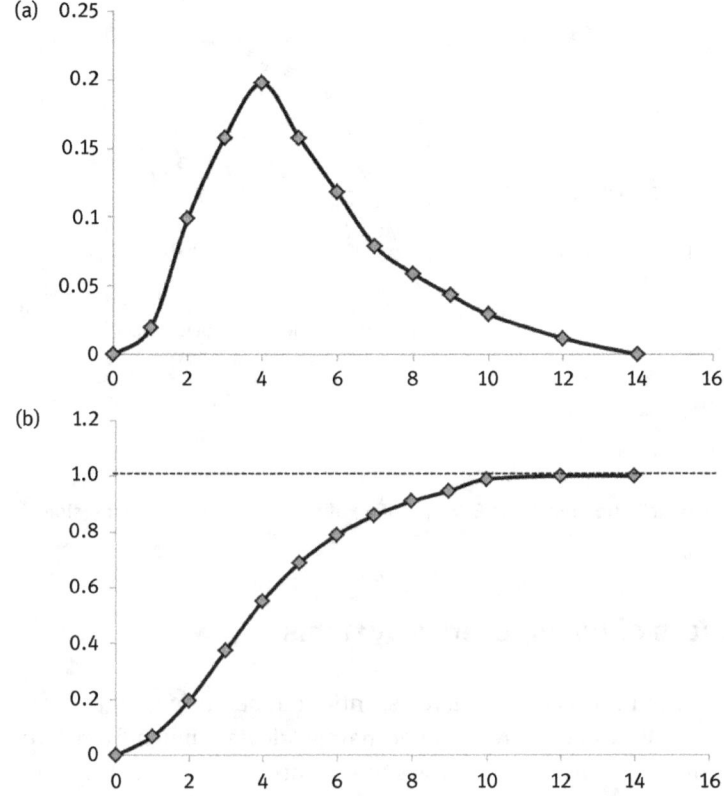

Figure 3.4: Pulse and step change experiments.

The impulse and step responses can be normalized, giving the residence time distribution functions $E(\theta)$ and $F(\theta)$ for the axial dispersion model ($\theta = t/\tau$):

$$F(\theta) = 1 - 4pe^p \sum_{i=1}^{\infty} \frac{(-1)^{i+1}\alpha_i^2}{(\alpha_i^2 + p^2)(\alpha_i^2 + p^2 + 2p)} e^{-\left(\theta\frac{(\alpha_i^2 + p^2)}{2p}\right)} \qquad (3.48)$$

$$E(\theta) = 2e^p \sum_{i=0}^{\infty} \frac{(-1)^{i+1}\alpha_i^2}{(\alpha_i^2 + p^2 + 2p)} e^{-\left(\theta\frac{(\alpha_i^2 + p^2)}{2p}\right)} \qquad (3.49)$$

where $p = Pe/2$ and α_i are the roots of transcendental functions (Salmi et al. (2011)).

It should be noted that the residence time distribution functions listed above are valid for the closed boundary conditions. For open boundary conditions, we get another solution which, in practice, does not differ very much from the solution obtained for closed boundary conditions (Hill & Root (2014)).

The residence time distribution functions obtained for different values of Péclet numbers are presented in Figure 3.5.

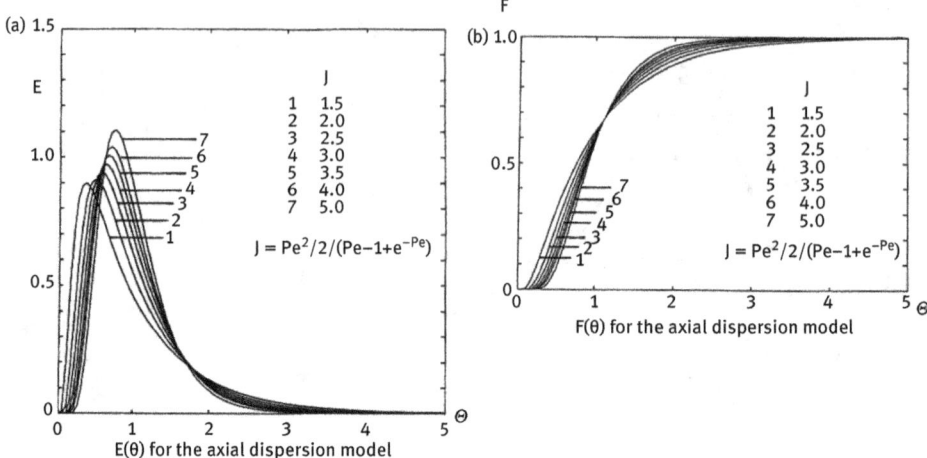

Figure 3.5: Residence time distribution functions $E(\theta)$ and $F(\theta)$ obtained for the axial dispersion model.

3.3 Energy balances of homogeneous systems

Energy balances are needed for reactors where essential temperature gradients appear; for instance, when the values of reaction enthalpies deviate much from zero. For exothermic reactions, the reaction enthalpies are negative and the temperature tends to increase in the reactor, and cooling is often necessary to keep the system under control, that is, to prevent the vaporisation of the reaction mixture, to suppress undesired side reactions and to prevent damage to the catalyst and reactor equipment. Thus, simulating the reactor temperature becomes crucially important to predict the system performance in a reliable way and to guarantee safe operating conditions. For endothermic systems, the reactor temperature tends to decrease, and it is important to predict in advance how much heat must be transferred to the reactor vessel to keep the process running.

The energy balances of ideal homogeneous reactors can be derived by considering the flow conditions of an infinitesimal volume element. The aim of this section is to illustrate the fact that the typical form of energy balance for a chemical reactor, where the reaction rates and reaction enthalpies are included, can be obtained from a very general form of the energy balance, including the flows of enthalpy, the internal energy and the heat exchange between the reactor and its surroundings. The general energy balance of a reactor volume element is illustrated in Figure 3.6.

We will start our consideration with an infinitely small volume element in an arbitrary homogeneous reactor, after which specific reactor configurations will be considered. A general energy balance of a volume element can be written as

3.3 Energy balances of homogeneous systems

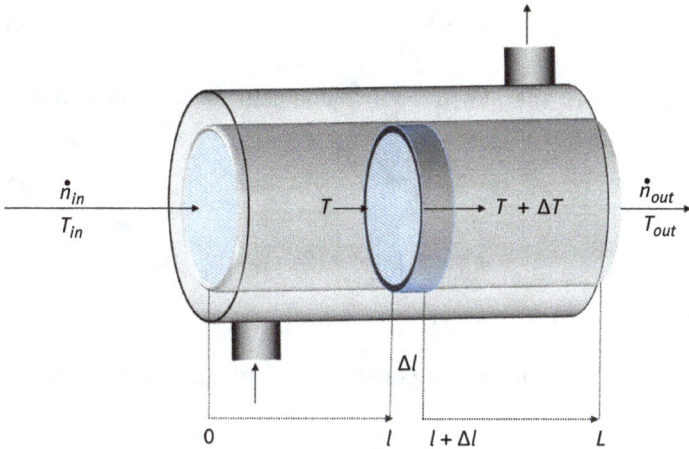

Figure 3.6: Energy balance for a reactor volume element.

$$\dot{H}_0 = \dot{H} + \Delta\dot{Q} + \frac{dU}{dt} \quad (3.50)$$

where $\dot{H}_0$ and $\dot{H}$ denote the inlet and outlet flow enthalpies, $\Delta\dot{Q}$ the amount of heat transferred away from the volume element for a time unit, and dU/dt the accumulation of internal energy. By using molar enthalpies, the enthalpy flows can be expressed with the molar flows. The mixing enthalpies are ignored below.

The use of molar enthalpies:

$$\dot{H} - \dot{H}_0 = \sum_i H_{mi}\dot{n}_i - \sum_i H_{mi0}\dot{n}_{i0} = \Delta\dot{H} = \Delta\sum_i H_{mi}\dot{n}_i \quad (3.51)$$

The term $\Delta\sum_i H_{mi}\dot{n}_i$ is expressed with the heat capacity (c_{pmi}), because the partial derivative $H_{mi}/dT = c_{pmi}$. Thus, we get

$$\Delta\sum_i H_{mi}\dot{n}_i = \sum_i \Delta H_{mi}\dot{n}_i + \sum_i H_{mi}\Delta\dot{n}_i \quad (3.52)$$

The question of what $\Delta\sum_i H_{mi}\dot{n}_i$ arises. The issue becomes clear as the molar heat capacity is introduced:

$$\sum_i \Delta H_{mi}\dot{n}_i = \sum_i \dot{n}_i c_{pmi}\Delta T \quad (3.53)$$

is valid, provided that the pressure effect is ignored.

For the second term in eq. (3.52), $\sum_i H_{mi}\Delta\dot{n}_i$, the mass balance gives

$$\Delta \dot{n}_i = r_i \Delta V - \frac{dn_i}{dt} \tag{3.54}$$

The generation rate is related to the stoichiometric coefficients and the reaction rates by

$$r_i = \sum_j \upsilon_{ij} R_j \tag{3.55}$$

and the molar amount being present in the volume element is

$$n_i = c_i \Delta V \tag{3.56}$$

Because the volume of the volume element is constant, the time derivatives are related by

$$\frac{dn_i}{dt} = \Delta V \frac{dc_i}{dt} \tag{3.57}$$

Thus, the term $\sum_i H_{mi} \Delta \dot{n}_i$ becomes

$$\sum_i H_{mi} \Delta \dot{n}_i = \sum_i H_{mi} \left(\sum_j \upsilon_{ij} R_j \Delta V - \frac{dc_i}{dt} \Delta V \right) = \sum_i \left(\sum_j \upsilon_{ij} H_{mi} \right) R_j \Delta V - \sum_i H_{mi} \frac{dc_i}{dt} \Delta V \tag{3.58}$$

where the sum $\sum_j \upsilon_{ij} H_{mi}$ de facto represents reaction enthalpy, ΔH_{rj}.

The term dU/dt can be elaborated further in an analogous manner:

$$\frac{dU}{dt} = \sum_i \frac{dU_i}{dt} = \sum_i \frac{d(U_{mi} n_i)}{dt} = \sum_i \frac{dU_{mi}}{dt} n_i + \sum_i \frac{dn_i}{dt} U_{mi}$$
$$= \left(\sum_i \frac{dU_{mi}}{dt} c_i + \sum_i \frac{dc_i}{dt} U_{mi} \right) \Delta V \tag{3.59}$$

The developed expressions (3.58) and (3.59) are inserted in the original balance eq. (3.50) giving

$$\sum_i \dot{n}_i c_{pmi} \Delta T + \sum_j \Delta H_{rj} R_j \Delta V - \sum_i H_{mi} \frac{dc_i}{dt} \Delta V + \sum_i \frac{dU_{mi}}{dt} c_i \Delta V$$
$$+ \sum_i U_{mi} \frac{dc_i}{dt} \Delta V + \Delta \dot{Q} = 0 \tag{3.60}$$

Thermodynamics gives us the well-known relations of heat capacities at a constant pressure (c_{pmi}) and constant volume (c_{vmi}):

$$\frac{dH_{mi}}{dT} = c_{pmi} \tag{3.61}$$

$$\frac{dU_{mi}}{dT} = C_{vmi} \qquad (3.62)$$

and the reactor temperature is included:

$$\frac{dU_{mi}}{dt} = \frac{dU_{mi}}{dT} \cdot \frac{dT}{dt} \qquad (3.63)$$

Combination of (3.62) and (3.63) gives

$$\frac{dU_{mi}}{dt} = C_{vmi} \frac{dT}{dt} \qquad (3.64)$$

Furthermore, the molar enthalpy and the molar internal energy are related by the basic thermodynamic relationship,

$$H_{mi} = U_{mi} + PV_{mi} \qquad (3.65)$$

where V_{mi} is the molar volume. Equation (3.65) implies that

$$(H_{mi} - U_{mi})\frac{dc_i}{dt}\Delta V = PV_{mi}\frac{dc_i}{dt}\Delta V \qquad (3.66)$$

The energy balance eq. (3.60) becomes:

$$\sum_i \dot{n}_i c_{pmi}\Delta T + \sum_j \Delta H_{rj} R_j \Delta V - P\sum_i V_{mi}\frac{dc_i}{dt}\Delta V + \sum_i c_{vmi}\frac{dT}{dt}c_i\Delta V + \Delta\dot{Q} = 0 \qquad (3.67)$$

By allowing the volume element to diminish, $\Delta V \to 0$, the differential equation is obtained:

$$\left(\sum_i c_{vmi} c_i\right)\frac{dT}{dt} - P\sum_i V_{mi}\frac{dc_i}{dt} + \left(\sum_i \dot{n}_i c_{pmi}\right)\frac{dT}{dV} + \sum_j \Delta H_{rj} R_j + \frac{d\dot{Q}}{dV} = 0 \qquad (3.68)$$

Equation (3.68) represents a very general form of energy balance and it can be applied to completely backmixed tank reactors and tube reactors with plug flow reactor (PFR). For tank reactors, a considerable simplification is possible, because the whole tank has a homogeneous content. This will be demonstrated in the following section. In most cases in practise, it is possible to ignore or simplify some of the terms in the general balance equation.

3.3.1 Tank reactor

Because of its homogeneous contents, the integration of eq. (3.68) over the entire tank volume can be carried out in a straightforward way,

$$\left(\sum_i c_{vmi} c_i\right) \frac{dT}{dt} \int_0^V dV - P \sum_i V_{mi} \frac{dc_i}{dt} \int_0^V dV + \int_{T_0}^T \left(\sum_i \dot{n}_i c_{pmi}\right) dT$$
$$+ \sum_j \Delta H_{rj} R_j \int_0^V dV + \int_0^{\dot{Q}} d\dot{Q} = 0 \qquad (3.69)$$

After division by the reactor volume, we obtain

$$\left(\sum_i c_{vmi} c_i\right) \frac{dT}{dt} - P \sum_i V_{mi} \frac{dc_i}{dt} + \frac{1}{V} \int_{T_0}^T \left(\sum_i \dot{n}_i c_{pmi}\right) dT + \sum_j \Delta H_{rj} R_j + \frac{\dot{Q}}{V} = 0 \quad (3.70)$$

in which the time derivative of the temperature is easily solved:

$$\frac{dT}{dt} = \frac{1}{\left(\sum_i c_{vmi} c_i\right)} \left(-\frac{1}{V} \int_{T_0}^T \left(\sum_i \dot{n}_i c_{pmi}\right) dT + \sum_j (-\Delta H_{rj}) R_j - \frac{\dot{Q}}{V} + P \sum_i V_{mi} \frac{dc_i}{dt}\right)$$
$$(3.71)$$

In a steady state, the time derivatives become zero,

$$\frac{dT}{dt} = 0, \qquad \frac{dc_i}{dt} = 0 \qquad (3.72)$$

and we get the algebraic equation

$$\frac{1}{V} \int_{T_0}^T \left(\sum_i \dot{n}_i c_{pmi}\right) dT = \sum_j (-\Delta H_{rj}) R_j - \frac{\dot{Q}}{V} \qquad (3.73)$$

Provided that if $\sum_i \dot{n}_i c_{pmi}$ is constant, eq. (3.73) can be rewritten as

$$\frac{T - T_0}{V} = \frac{1}{\sum_i \dot{n}_i c_{pmi}} \left(\sum_j (-\Delta H_{rj}) R_j - \frac{\dot{Q}}{V}\right) \qquad (3.74)$$

Generally, the molar-based heat capacities and molar flows can be related to the corresponding mass-based quantities by the well-known relation.

$$\sum_i \dot{n}_i c_{pmi} = \dot{m}_i c_p \qquad (3.75)$$

After introducing eq. (3.75) into eq. (3.74), we get

$$\frac{T - T_0}{V} = \frac{1}{\dot{m}_i c_p} \left(\sum_j (-\Delta H_{rj}) R_j - \frac{\dot{Q}}{V}\right) \qquad (3.76)$$

We should keep in mind that eqs. (3.74) and (3.76) are not general: if $\sum_i \dot{n}_i c_{pmi}$ changes considerably, the integration from the inlet to the outlet temperature is carried out as in eq. (3.73). In addition, the term $P \sum V_{mi} dc_i/dt$ in eq. (3.71) is often ignored for liquid-phase reactions.

The heat transfer term ($\dot{Q}$) is expressed through the temperature difference between the reactor (T) and the surroundings (T_c). S is the heat transfer surface area and U denotes the overall heat transfer coefficient, including heat transfer resistances in the fluid film inside the reactor, the reactor wall as well as resistance on the outer surface of the reactor wall.

$$\dot{Q} = U \cdot S \cdot (T - T_c) \tag{3.77}$$

3.3.2 Tubular plug flow reactor

For tubular plug flow reactors, the general energy balance eq. (3.68) is directly valid:

$$\left(\sum_i c_{vmi} c_i \right) \frac{dT}{dt} - P \sum_i V_{mi} \frac{dc_i}{dt} + \left(\sum_i \dot{n}_i c_{pmi} \right) \frac{dT}{dV} + \sum_j \Delta H_{rj} R_j + \frac{d\dot{Q}}{dV} = 0 \tag{3.78}$$

For reactors operating in a steady state, the time derivatives are zero: $dT/dt = 0$, $dc_i/dt = 0$ giving

$$\frac{dT}{dV} = \frac{1}{\sum_i \dot{n}_i c_{pmi}} \left(\sum_j (-\Delta H_{rj}) R_j - \frac{d\dot{Q}}{dV} \right) \tag{3.79}$$

The use of mass-based heat capacity $\sum_i \dot{n}_i c_{pmi} = \dot{m}_i c_p$ leads to

$$\frac{dT}{dV} = \frac{1}{\dot{m}_i c_p} \left(\sum_j (-\Delta H_{rj}) R_j - \frac{d\dot{Q}}{dV} \right) \tag{3.80}$$

For cylindrical tubular reactors, $d\dot{Q}/dV$ can be regarded as a constant because of geometry:

$$\frac{d\dot{Q}}{dV} = U(T - T_j) \frac{dS}{dV} = U \frac{S}{V_R} (T - T_j) \tag{3.81}$$

where S is the total heat transfer area. For cylindrical tubes, the ratio is obtained from the basic geometry:

$$\frac{S}{V_R} = \frac{2\pi R_T L_T}{\pi R_T^2 L_T} = \frac{4}{D_T} \tag{3.82}$$

where R_T, D_T and L_T denote the radius, the diameter and the length of the tube. The temperature outside the reactor is denoted by

$$T_j = T_c \tag{3.83}$$

3.3.3 Batch reactor

For the ideal completely backmixed batch reactor, the energy balance is written as

$$\dot{Q} + \frac{dU}{dt} = 0 \tag{3.84}$$

Furthermore, for the mass, balance is valid

$$\frac{dn_i}{dt} = V \sum_j v_{ij} R_j \tag{3.85}$$

Introduction of molar quantities in eq. (3.84) gives

$$\dot{Q} + \frac{d(\sum_i U_{mi} n_i)}{dt} = 0 \tag{3.86}$$

Provided that the reactor volume is constant, differentiation of eq. (3.86) leads to,

$$\dot{Q} + \sum_i \frac{dU_{mi}}{dt} c_i V + \sum_i U_{mi} V \frac{dc_i}{dt} = 0 \tag{3.87}$$

where the concentration derivative is obtained from eq. (3.85),

$$\frac{dn_i}{dt} = V \frac{dc_i}{dt} = V \sum_j v_{ij} R_j \tag{3.88}$$

The derivative of internal energy is obtained from the molar heat capacity as follows,

$$\frac{dU_{mi}}{dt} = \frac{dU_{mi}}{dT} \frac{dT}{dt} = c_{vmi} \frac{dT}{dt} \tag{3.89}$$

Equations (3.88) and (3.89) are inserted in the balance eq. (3.87). The result becomes

$$\dot{Q} + \left(\sum_i c_{vmi} c_i\right) V \frac{dT}{dt} + \sum_j v_{ij} R_j \sum_i U_{mi} V = 0 \tag{3.90}$$

After rearrangements, we get eq. (3.91), in which the temperature derivative is solved explicitly:

$$\left(\sum_i c_{vmi} c_i\right) \frac{dT}{dt} = \sum_j (-\Delta U_{rj}) R_j - \frac{\dot{Q}}{V} \tag{3.91}$$

If the change in internal energy is approximated by the change in reaction enthalpy, we obtain

$$\frac{dT}{dt} = \frac{1}{\sum_i c_{vmi} c_i} \left(\sum_j (-\Delta H_{rj}) R_j - \frac{\dot{Q}}{V}\right) \tag{3.92}$$

Batch reactors are quite frequently used for liquid-phase processes, in which the difference between internal energy change (ΔU_{rj}) and enthalpy change (ΔH_{rj}) is negligible, $\Delta U_{rj} \approx \Delta H_{rj}$.

We can thus simplify

$$\sum_i c_{vmi} c_i = \sum_i \frac{c_{vmi} n_i}{V_R} = \frac{m c_v}{V_R} = \rho_0 c_v \approx \rho_0 c_p \tag{3.93}$$

where the product $\rho_0 c_p$ is the heat capacity of the reacting liquid. The energy balance (3.92) becomes

$$\frac{dT}{dt} = \frac{1}{\rho_0 c_p} \left(\sum_j (-\Delta H_{rj}) R_j\right) - \frac{\dot{Q}}{V} \tag{3.94}$$

Even here, the heat flux ($\dot{Q}$) is given by the well-known equation,

$$\dot{Q} = U \cdot S \cdot (T - T_c) \tag{3.95}$$

Equation (3.94) is a very practical form of batch reactor energy balance, applicable to numerous liquid-phase processes.

3.3.4 Semi-batch reactors

For SBRs, different ways of operation exist, but the most typical one is feeding some of the components into a reacting liquid. An operational form for the energy balance of a liquid-phase SBR is obtained from the general balance eq. (3.71) valid for a tank reactor. Typically, mass-based heat capacities are used and c_{vmi}, c_{pmi}. In addition, the term $P \sum V_{mi} dc_i/dt$ is ignored. The following form of the energy balance equation is in most cases useful for liquid-phase SBRs:

$$\frac{dT}{dt} = \frac{1}{\rho_0 c_p} \left(-\frac{\rho_0}{\tau} \int_{T_0}^{T} c_p dT + \sum_j (-\Delta H_{rj}) R_j - \frac{\dot{Q}}{V_R}\right) \tag{3.96}$$

where $\dot{Q}$ is given by eq. (3.95).

For correlations of heat transfer coefficient U, the reader is referred to literature dedicated to this topic (e.g. Froment et al. (2011)). In case of more sophisticated arrangements for heat transfer, for example, where internal cooling coils are used, additional heat transfer terms are included in the energy balance. In the discussion above, heat transfer from the reactor towards its surroundings was considered (cooling of the reactor). The treatment is general, however, in case of heating of the reactor vessel: $T_0 > T$ and $\dot{Q}$ is negative. The balance equations presented are valid in this case, too. The energy balance is seldom used in its most general cases, but simplifications are introduced or some special cases are considered.

3.4 Physical properties and correlations of homogeneous systems

The most essential physical properties of homogeneous systems and correlation equations for homogeneous reactors are discussed briefly in this section. For predictive models, the reader is referred, for example, to the book by Reid et al. (1988), which also includes an extensive data bank of physical properties.

3.4.1 Heat capacity and reaction enthalpy

Molar heat capacities at constant pressure (c_{pmi}) are related to the overall mass-based heat capacity (c_p) of the system by eqs. (3.97) and (3.98) valid for continuous and discontinuous systems,

$$c_p m = \sum_i c_{pmi} n_i \qquad (3.97)$$

$$c_p \dot{m} = \sum_i c_{pmi} \dot{n}_i \qquad (3.98)$$

Completely analogous relations are valid for heat capacities at constant volume (c_{vmi} and c_v), but they are not discussed in detail here.

After recalling the relation $n_i = x_i \cdot n$; $M = m/n$ average molar mass, and n is the total amount of substance, we get

$$c_p \bar{M} = \sum_i x_i c_{pmi} = \overline{c_{pm}} \qquad (3.99)$$

where $\bar{M} = \sum x_i M_i$. For ideal gases, the heat capacities are related by

$$c_{pmi} - c_{vmi} = R \qquad (3.100)$$

where R is the universal gas constant. For liquid-phase systems, the difference between c_{pmi} and c_{vmi} is typically negligible.

The reaction enthalpies are updated with the formula

$$\Delta H_{rj}(T) = \Delta H_{rj}(T_{ref}) + \int_{T_{ref}}^{T} \sum_{i=1}^{N} v_{ij} c_{pmi} dT \tag{3.101}$$

It should be noted that eq. (3.101) does not account for phase transitions between temperatures T_{ref} and T. The reaction enthalpy at the reference temperature (T_{ref}) is obtained from the enthalpies of formation (H_i^f) as expressed in eq. (3.102). Enthalpies of formation at reference temperatures are listed, for example, by Reid et al. (1988). For liquid-phase systems, a simplified treatment is often applied, $\Delta H_r \approx$ constant, where ΔH_r is obtained from calorimetric measurements.

$$\Delta H_{rj}(T_{ref}) = \sum_{i=1}^{N} v_{ij} H_i^f \tag{3.102}$$

3.4.2 Pressure drop in tubular reactors

Pressure drop in homogeneous tube reactors is typically small compared to the pressure drop in other equipment, that is, pipelines for transporting gases and liquids to and from the factory, as well as pipelines connecting the reactors to separation equipment. Fanning equation (Froment et al. (2011)) is adequate for pressure drop calculations in tube reactors,

$$-\frac{dP}{dl} = 2f\frac{\rho w^2}{d_T} + \rho w \frac{dw}{dl} \tag{3.103}$$

where w is the superficial velocity ($w = \dot{V}/A$) and f is the friction factor. For a laminar flow, the friction factor is obtained from a theoretical relationship

$$f = \frac{16}{Re} \tag{3.104}$$

whereas an empirical relation

$$f = 0.046 Re^{-0.2} \tag{3.105}$$

has been proposed for a turbulent flow, that is, for Reynolds numbers exceeding 2500 (Froment et al. (2011)).

The pressure drop is in principle coupled to mass and energy balances, especially to the energy balance. In fact, this connection is rather loose, because the factor ρw is constant throughout the reactor. The superficial velocity of ideal gases is expressed by $w = GRT/(\bar{M}P)$, where $G = \dot{m}/A$ and $\bar{M}$ is the average molar mass. Thus, the main connection is with the energy balance, since the superficial velocity is proportional to temperature (T).

3.4.3 Dispersion coefficient

The best way to obtain the numerical value of the axial dispersion coefficient is to carry out pulse or step change experiments with an inert tracer. From experimentally recorded $E(\Theta)$ and $F(\Theta)$ curves, the variance of the signal can be calculated as follows (Figure 3.5):

$$\sigma_\Theta^2 = \int_0^\infty (\Theta - 1)^2 E(\Theta) d\Theta = \int_0^\infty \Theta^2 E(\Theta) d\Theta - 1 \tag{3.106}$$

where $\Theta = t/\bar{t}$. For the axial dispersion model, the following relationship is valid between the variance and the Péclet number

$$\sigma_\Theta^2 = \frac{2}{Pe^2}\left(Pe - 1 + e^{-Pe}\right) \tag{3.107}$$

Thus, after obtaining the variance, the Péclet number can be calculated iteratively from eq. (3.107).

In case of tubular reactors operated on an industrial scale and design projects, tracer experiments are usually not possible, but the Péclet number can be estimated from the available correlations in literature. Laminar and turbulent flow patterns are treated separately.

For laminar flow ($Re < 2000$), the following equation has been derived theoretically:

$$D = D_m + \frac{w^2 d^2}{192 D_m}, \quad 1 < Re < 2000 \tag{3.108}$$

where D_m is the molecular diffusion coefficient, and w and d denote superficial velocity and tube diameter, respectively. Equation (3.108) can be rewritten using two dimensionless quantities, namely the Reynolds (Re) and Schmidt (Sc) numbers,

$$Re = \frac{wd}{\upsilon} \text{ and } Sc = \frac{\upsilon}{D_m} \tag{3.109}$$

The correlation now becomes

$$\frac{1}{Pe_R} = \frac{1}{Re\,Sc} + \frac{Re\,Sc}{192} \tag{3.110}$$

where Pe_R is the radial Péclet number defined as

$$Pe_R = \frac{wd}{D} \qquad (3.111)$$

where d is the tube diameter. For turbulent flows ($Re > 2000$), the following empirical correlation has been proposed:

$$\frac{1}{Pe_R} = \frac{a}{Re^\alpha} + \frac{b}{Re^\beta}, \quad Re > 2000 \qquad (3.112)$$

where $a = 3 \cdot 10^7$, $b = 1.35$, $\alpha = 2.1$ and $\beta = 1/8$.

Equation (3.110) predicts that the Péclet number for a laminar flow has a maximum at $ReSc = 192^{1/2}$, while the model for a turbulent flow gives a monotonously increasing Péclet number as a function of the Reynolds number.

The longitudinal Péclet number (Pe) needed in the axial dispersion model is obtained from the simple relationship

$$Pe = Pe_R \frac{L}{d} \qquad (3.113)$$

where L is the reactor length.

3.5 Numerical solution of homogeneous reactor models

Analytical solution of homogeneous reactor models is possible in simple cases, for instance when the system is isothermal and all of the rate equations are linear with respect to component concentrations. Furthermore, in isothermal systems where single reactions only take place, analytical and semi-analytical solutions can sometimes be obtained. The classical approach based on analytical solutions is presented in numerous textbooks (e.g. Levenspiel (1999), Froment et al. (2011), Salmi et al. (2011)). For non-isothermal systems, numerical solution is the only way, since the rate constants have an exponential coupling to the temperature according to the law of Arrhenius. We will here consider the general methodology for solving reactor models through numerical treatment.

3.5.1 Model structures and algorithms

The mathematical structures of the homogeneous reactor models presented in this Chapter are summarized in Table 3.1. Most of the models are ODEs of the initial value type (IVP). Even the non-steady-state (dynamic) model for the PFR can easily be converted into an ODE, by applying discretization with respect to the space derivative $dy/dz = [d\underline{c}/dz dT/dz]$. Backward differences are used to approximate the derivative. The approach is called the method of lines (Schiesser (1991)). It is

Table 3.1: Mathematical structures of homogeneous reactor models.

Model	Mathematical problem structure	
Steady-state CSTR	$f(y) = 0$	NLE
Dynamic CSTR	$\dfrac{dy}{dx} = f(y)$	ODE(IVP)
Steady-state PFR	$\dfrac{dy}{dz} = f(y)$	ODE(IVP)
Dynamic PFR	$\dfrac{\partial y}{\partial t} = -A \cdot \dfrac{\partial y}{\partial z} + f(y)$	PDE(IVP)
Batch reactor (BR) Semi-batch reactor (SBR)	$\dfrac{dy}{dt} = f(y)$	ODE(IVP)
Steady-state axial dispersion model (ADM)	$\dfrac{d^2 y}{dz^2} + B\dfrac{dy}{dz} + f(y) = 0$	ODE(BVP)
Dynamic axial dispersion model (ADM)	$\dfrac{\partial y}{\partial t} = A\dfrac{\partial^2 y}{\partial z^2} + B\dfrac{\partial y}{\partial z} + f(y)$	PDE
IVP = initial value problem	NDE = nonlinear differential equation	
BVP = boundary value problem	ODE = ordinary differential equation	
	PDE = partial differential equation	

important to use backward differences for the derivatives dy/dz originating from plug flow; a wrong numerically oscillating solution is obtained if central differences are applied. Backward differences of different orders are obtained from Newton's backward interpolation polynomial. The weights of backward difference formulae of various orders are summarized in Romanainen and Salmi (1994). The simplest two-point backward difference formula is

$$\left(\frac{dy}{dz}\right)_{y=y_0} = \frac{1}{\Delta z}(y_0 - y_1) \qquad (3.114)$$

and the three-point formula is written in the form

$$\left(\frac{dy}{dz}\right)_{y=y_i} = \frac{1}{\Delta z}\left(\frac{3}{2}y_i - \frac{4}{2}y_{i-1} + \frac{1}{2}y_{i-2}\right) \qquad (3.115)$$

The use of a higher order difference formula is analogous. The accuracy of the approximation improves as the order of the difference formula increases.

Some backward difference formulae for first derivatives originating from plug flow are presented as follows

$$\left(\frac{dy}{dz}\right)_{y=y_i} = \frac{1}{12h}(25y_i - 48y_{i-1} + 36y_{i-2} - 16y_{i-3} + 34y_{i-4}) \quad (3.116)$$

$$\left(\frac{dy}{dz}\right)_{y=y_i} = \frac{1}{60h}(137y_i - 300y_{i-1} + 300y_{i-2} - 200y_{i-3} + 75y_{i-4} + 12y_{i-5}) \quad (3.117)$$

where $h = \Delta z$. For terms originating from dispersion or diffusion, central differences should be used. Some central difference formulae are listed below.

$$\left(\frac{d^2y}{dz^2}\right)_{y=y_i} = \frac{1}{180h^2}(2y_{i-3} - 27y_{i-2} + 270y_{i-1} - 490y_i + 270y_{i+1} - 27y_{i+2} + 2y_{i+3}) \quad (3.118)$$

$$\left(\frac{d^2y}{dz^2}\right)_{y=y_i} = \frac{1}{5040h^2}(-9y_{i-4} + 128y_{i-3} - 1008y_{i-2} + 8064y_{i-1} - 14350y_i + 8064y_{i+1}$$
$$- 1008y_{i+2} + 128y_{i+3} - 9y_{i+4})$$
$$(3.119)$$

An important phenomenon which should be kept in mind is the solution of the dynamic PFR model with the method of lines: the derivatives in the numerical solution tend to smoothen the concentration fronts progressing inside the reactor. The observed effect is similar to real axial dispersion, but in fact it is an artefact only. This problem is known as numerical diffusion and it is worst for slow reactions, for which the reactant step response is close to a step function. The effect is illustrated in Figure 3.7. If the problem is more severe, more sophisticated techniques, such as adaptive grid methods, should be applied.

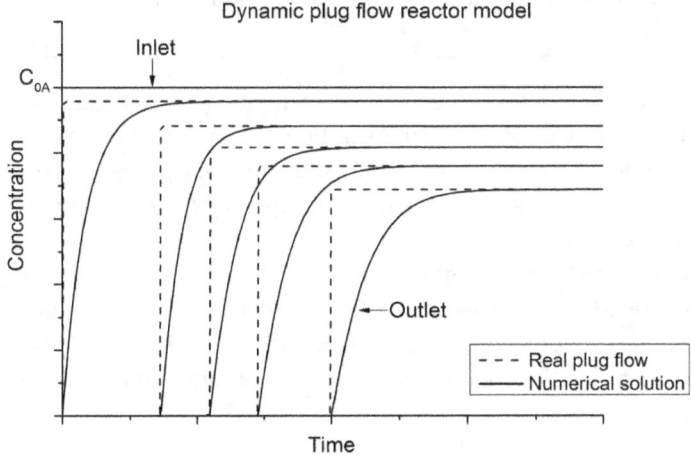

Figure 3.7: Illustration of numerical diffusion in the solution of the dynamic plug flow model.

For initial value problem ODEs, a number of numerical algorithms have been developed. Some of them work very well for reaction engineering problems, while others should be avoided. A characteristic feature of problems in chemical kinetics and reactors is the risk of stiffness of the system. Stiffness is a concept used by mathematicians to characterize the time constants of systems of ODEs, and an exact definition of stiffness is based on the eigenvalues of the system. In practise, a detailed analysis of stiffness is not carried out, but the parameters of the system can give a feel of it. For instance, let us assume that we have a chemical system in which some of the rate constants have very low values, while others are very high. This implies that some of the reaction steps might be close to their equilibria, while other steps progress very slowly. This kind of a system is stiff with a high probability. Furthermore, discretization of the spatial coordinate (in solution of the dynamic PFR model) increases the stiffness of the system. This is easy to understand intuitively, since different reactions are active in different parts of the reactor tube.

The well-known explicit methods, such as the explicit Euler's method and explicit Runge-Kutta methods of different orders, are highly inefficient in the solution of stiff ODEs. They suffer from serious stability problems, and an impractically small step length should be used to achieve a stable solution. Implicit methods, on the other hand, show good stability properties and are to be preferred in computations in reaction engineering. Two kinds of methods for stiff ODEs have been developed extensively in recent decades, namely linear multistep methods, semi-implicit Runge-Kutta methods and public domain computer codes have been devised (Hindmarsh (1983), Kaps and Wanner (1981)). The algorithms are even available in high-level programming languages, such as Matlab and Mathematica.

The simplest method showing a good stability for stiff ODEs is the implicit Euler method. The algorithm can be described as follows for the solution of $dy/dx = f(y)$,

$$y_n = y_{n-1} + \Delta x \cdot f(y_n) \tag{3.120}$$

where Δx is the discretisation interval. The method is illustrated in Figure 3.8. Observe that the algorithm normally leads to the solution of non-linear equations, since $f(y_n)$ is typically a non-linear function of y_n. The linear multistep methods and semi-implicit Runge-Kutta methods are treated in detail in Appendix A.2, where more advanced algorithms are presented.

The steady-state CSTR model represents non-linear algebraic equations (NLEs). Linear equations are obtained only for isothermal first-order reactions. For non-linear equations, the classical method of numerical solution is the algorithm of Newton and Raphson, which for a system of a single non-linear equation $f(y) = 0$ is given as

$$y_{k+1} = y_k - \frac{f(y_k)}{f'(y_k)} \tag{3.121}$$

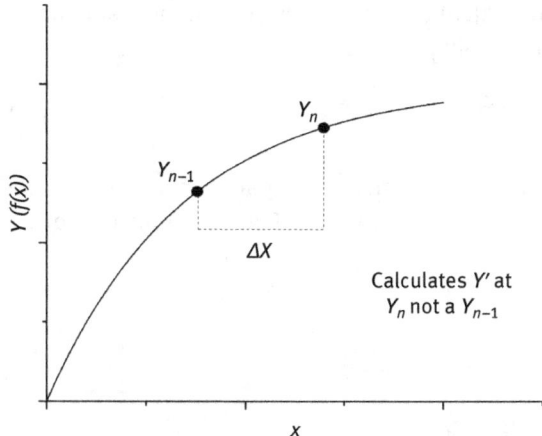

Figure 3.8: The principle of the implicit Euler method.

where k is the iteration index. The method is extended to systems of several unknowns in a straightforward manner: the derivative $f'(y_k)$ is replaced by the Jacobian containing all of the partial derivatives $\partial f_i/\partial y_j$. The details are explained in Appendix A.1.

The advantage of the Newton-Raphson method is that it is rapid – typically a quadratic convergence is achieved, provided that the initial guess of the solution is close to the actual solution. How to obtain a reasonable initial guess? Real experimental data may help in some cases, but this approach is not a general one. Usually, the principle of continuity is very efficient: if the solution is known for one parameter value, the solution for another parameter value (x) can be obtained by proceeding towards the new parameter value stepwise and by using the solution obtained for the previous parameter value as an initial guess:

$$y_0(x + \Delta x) = y(x) \tag{3.122}$$

A natural selection for the continuity parameter is the reactor volume or space time. For instance, for isothermal CSTRs, we have the mass balance equations

$$\frac{c_i - c_{0i}}{\tau} = r_i, \quad \tau = \frac{V}{V_0} \tag{3.123}$$

$$f(c_i) = c_i - c_{0i} - r_i \tau = 0 \tag{3.124}$$

If the solution for a fixed value of τ is known, we can calculate the solution for $\tau + \Delta\tau$ by using the solution at τ as an initial estimate. Such a calculation provides very useful information from the reaction engineering viewpoint: we obtain information about to what extent the reactor performance changes with respect to space time (or reactor volume). The calculation provides information similar to what is obtained from the solution of the plug flow model,

$$\frac{dc_i}{d\tau} = r_i \qquad (3.125)$$

for which the solution of the ODEs gives c_i as a function of space time. The performances of ideal reactors can be compared as illustrated in Figure 3.9 for first-order kinetics.

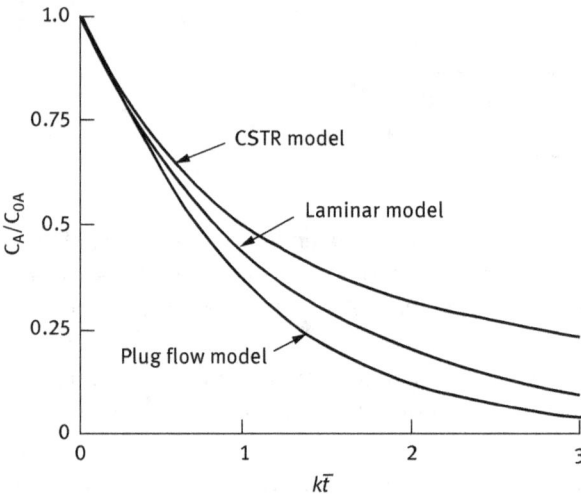

Figure 3.9: Comparison of the performances of ideal reactors (plug flow, laminar and CSTR).

In the worst case, if the convergence of the Newton-Raphson model for a steady-state CSTR is not achieved, it is always possible to switch to the dynamic CSTR model and solve it by the algorithms designed for stiff ODEs. The numerical solution is continued until a steady state is attained, in practise, that is, until the criterion $|dy/dt| < \varepsilon_i$ (ε_i close to 0) (y is concentration or molar amount) is fulfilled for each component concentration (or molar flow) and reactor temperature.

3.5.2 Software build-up

In the build-up of numerical simulation software for reactor models, several possibilities and options exist. How to do it in practise is very much a matter of taste and dependent on the character of the research and development project. You can always say, "I did it my way," but some ways produce more enduring value than

others. A good principle is to separate the different tasks in simulation. The following tasks typically exist in each case:
- input of chemical engineering data (initial concentrations, temperatures, reactor dimensions and kinetic, thermodynamic as well as mass and heat transfer parameters)
- input of data for steering the numerical solution (numerical methods, step-length selection, convergence criteria, etc.)
- definition of kinetic and thermodynamic models (reaction rates, calculation of rate and equilibrium constants)
- definition of mass and heat transfer models (e.g. correlations for diffusion and dispersion coefficients, heat and mass transfer coefficients and areas)
- definition of mass and energy balances along with equations for pressure drop
- definition of partial derivatives needed for the model solution
- numerical solver for the differential and/or algebraic equations involved in the model
- output routines, which not only reveal the results (e.g. concentration, temperature and pressure profiles) but also give information about the success or failure of the numerical solution process

A typical simulation program structure is illustrated schematically in Figure 3.10. For many cases of homogeneous reactions, a simplified program structure is sufficient as shown in Figure 3.11.

The key issue is to avoid mixing up the different tasks. For instance, the ideal gas law should preferably not be inbuilt in the model subroutine but assigned a separate routine: this way, the door is kept open for replacing the ideal gas law by more sophisticated gas laws in future applications.

Calculation of reaction rates is a well-defined subtask which should be carried out separately and not mixed with the mass balance equations of components in the reactor. This way, kinetic models can be transferred to other reactor models, even if completely different simulators are used. Perhaps the most important thing is to keep the numerical solution algorithm strictly separated from chemical engineering subroutines. Algorithms, such as semi-implicit Runge-Kutta methods, are completely independent in this sense from the actual chemical engineering application. As numerical mathematics makes new progress, an old ODE solver can easily be replaced by a new, improved one, provided that the numerical subroutines in the software have not been mixed up with the chemical engineering ones.

The crucial issue is, of course, who should write the numerical solver – the chemical engineer or the numerical mathematician. Nowadays the answer is rather unequivocal; namely the latter. Several excellent numerical codes for solving NLEs and ODEs exist as summarized in Appendices A-B. These should be tried first and

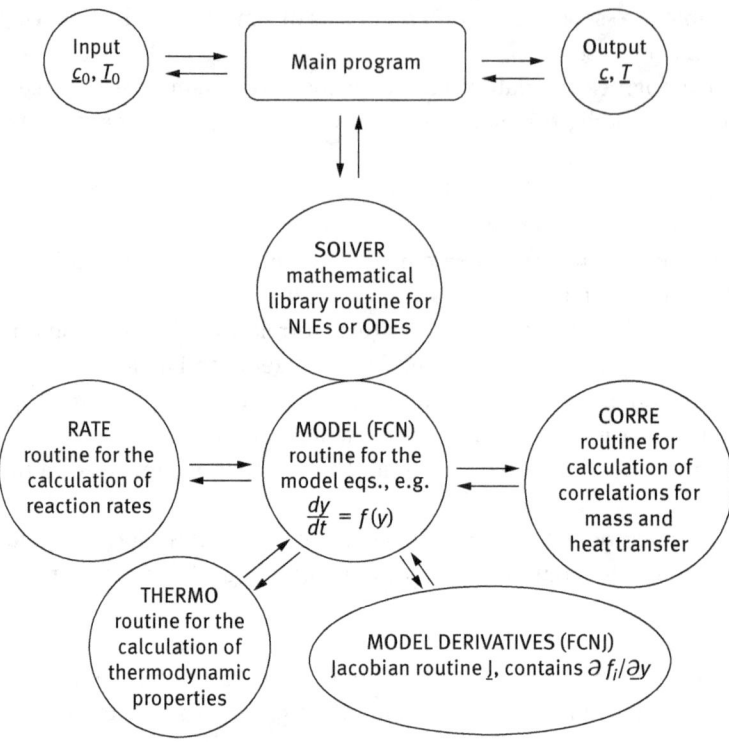

Figure 3.10: Simulation program structure.

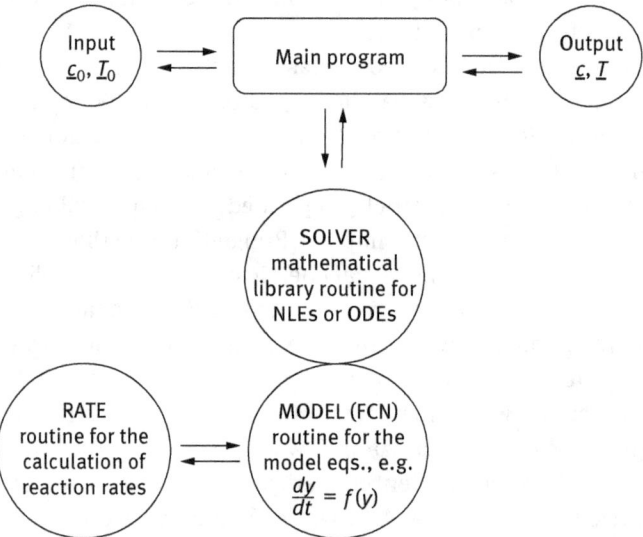

Figure 3.11: Simplified simulation program structure.

applied in a straightforward manner to routine problems. Only for extremely difficult cases, or in cases where a completely new idea has been invented for numerical solution, a specific numerical effort should be made by the chemical engineer! The situation was different in the past, but nowadays, very few of us assemble our own cars by ourselves!

4 Modelling of fixed beds and fluidized beds

Catalysts are chemical compounds which are able to accelerate chemical processes without being consumed in them. A catalyst has a profound enhancing function in a chemical process. Since the discovery of the catalytic effect by the world-famous Swedish chemist J.J. Berzelius in 1835, catalysis has had an enormous breakthrough in the chemical industry: about 90% of chemicals of today are produced with the aid of catalysis. If the catalyst is soluble in the reaction medium, we talk about homogeneous catalysis. If the catalyst forms a separate phase, typically a separate solid phase, the process is called heterogeneous catalysis. It is hard to imagine modern chemical industry and consumer society without catalysis. Catalysis contributes to the production of fuels, ammonia and sulphuric acid, fertilizers, polymers, plastics as well as many pharmaceuticals and fine chemicals. A short overview of contemporary catalytic processes is provided in Table 4.1. The discussion shows the central role of catalytic processes and reactors in chemical reaction engineering.

Table 4.1: Some catalytic processes applied on an industrial scale.

Chemical bulk industry	Oil refining	Petrochemical industry
Steam reforming	Reforming	Ethene oxide
Carbon monoxide conversion	Isomerization	Ethene dichloride
Carbon monoxide methanization	Dehydrogenation	Vinyl acetate
Methanol synthesis	Desulphurization	Butadiene
Oxo-synthesis	Hydrocracking	Maleic acid anhydride
		Phtalic acid anhydride
		Cyclohexane

This Chapter is devoted to catalytic two-phase reactors where solid catalyst particles and a gas or liquid phase are present. The physical configurations of catalytic reactors are numerous and under continuous development. However, two main categories are dominant: catalytic fixed beds and fluidized beds. In general, fixed beds are characterized by stagnant, relatively large catalyst particles (in mm-cm scale), while fluidized beds operate with moving, very small catalyst particles (diameters in micrometre scale) (Figure 4.1).

In any catalytic systems, chemical reactions interact with mass and heat transfer effects. For example, mass and heat transfer effects are present inside porous catalyst particles (Figure 4.2) as well as in the surrounding fluid films. In addition, heat transfer from and to the catalytic reactor makes an essential contribution to the energy balance. At the core of modelling a two-phase catalytic reactor is the catalyst particle, namely simultaneous reaction and diffusion in the pores of the particle. These effects

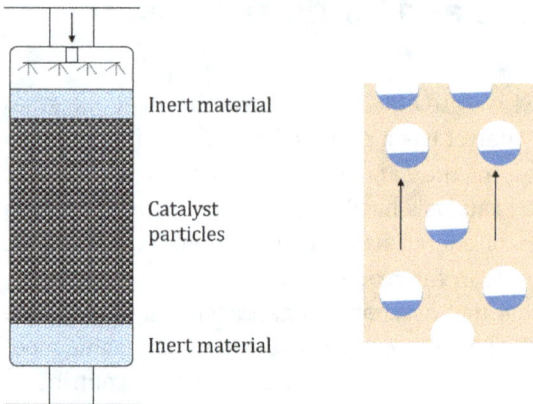

Figure 4.1: A catalytic fixed bed and fluidized bed.

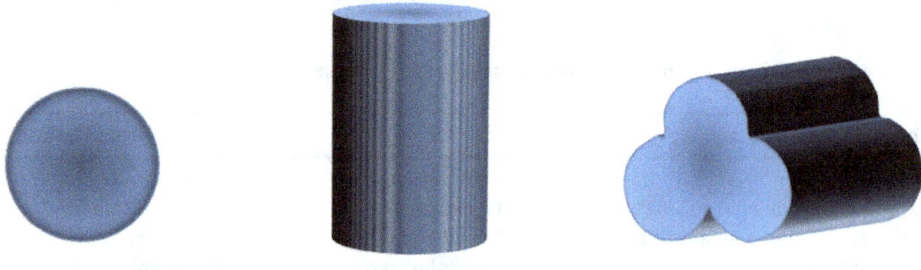

Figure 4.2: Industrial catalyst particles.

are completely analogous with the reaction-diffusion effects in liquid films appearing in gas-liquid systems. Thus, the formulae presented in the next section are not only valid for catalytic reactions but also for gas-liquid processes (Chapter 6).

4.1 Simultaneous reaction and diffusion in fluid films and porous media

A general continuity equation (mass balance) for a component in a layer where the transport proceeds through diffusion and chemical reactions take place simultaneously is written for an infinitesimal volume element,

$$(N_i A)_{in} + r_i a \Delta V = (N_i A)_{out} + \frac{dn_i}{dt} \tag{4.1}$$

where N_i and r_i denote the flux and the generation rate, respectively. A is the cross-section area of the flux, ΔV is the volume element and n_i is the amount of substance in the volume element. The factor $\alpha = 1$ for homogeneous kinetics, but equals the catalyst density ($\alpha = \rho_p$) for catalytic systems. The amount of substance in the volume element, n_i, is expressed through the concentration (c_i) and volume of the fluid phase in the volume element ($\varepsilon \Delta V$). The factor (ε) is 1 for homogeneous systems but <1 for porous catalysts, for which it denotes the catalyst porosity.

By denoting the difference

$$\Delta(N_i A) = (N_i A)_{out} - (N_i A)_{in} \tag{4.2}$$

we get

$$-\Delta(N_i A) + r_i \alpha \Delta V = \varepsilon \Delta V \frac{dc_i}{dt} \tag{4.3}$$

Division by the volume element ΔV and allowing $\Delta V \to 0$ gives

$$\varepsilon \frac{dc_i}{dt} = -\frac{d(N_i A)}{dV} + r_i \alpha \tag{4.4}$$

This is a very general form of the transport equation, and it can be applied to any stagnant layer in which the overall fluid velocity is zero. Further treatment depends on the models for geometry and diffusion used. The most common and simple geometry is a slab (solid or liquid films and catalyst particles in the form of corn flakes): $dV = A dr$, where dr is the thickness element of the catalyst.

In this special case, we obtain

$$\varepsilon \frac{dc_i}{dt} = -\frac{d(N_i)}{dr} + r_i \alpha \tag{4.5}$$

For a spherical geometry, on the other hand, $A = 4\pi r^2$ and $dV = 4\pi r^2 dr$ giving the relationship

$$\varepsilon \frac{dc_i}{dt} = -\frac{d(N_i r^2)}{r^2 dr} + r_i \alpha \tag{4.6}$$

that is,

$$\varepsilon \frac{dc_i}{dt} = -\frac{dN_i}{dr} + \frac{2}{r} N_i + r_i \alpha \tag{4.7}$$

Analogously it can be shown for the third ideal geometry, an infinitely long cylinder, that the balance equation becomes

$$\varepsilon \frac{dc_i}{dt} = -\frac{d(N_i r)}{r dr} + r_i \alpha \tag{4.8}$$

Suggesting that a general formulation for the reaction-diffusion equation is

$$\varepsilon \frac{dc_i}{dt} = -\frac{d(N_i r^s)}{r^s dr} + r_i \alpha \qquad (4.9)$$

that is,

$$\varepsilon \frac{dc_i}{dt} = -\frac{dN_i}{dr} + \frac{s}{r} N_i + r_i \alpha \qquad (4.10)$$

where $s = 0$ for slabs, $s = 1$ for infinitely long cylinders and $s = 2$ for spheres. All real systems can be placed within the interval [0, 2] as will be shown in the treatment of porous catalysts (Chapters 4–5) and gas-liquid systems (Chapter 6). It should be emphasized that non-integer numbers are possible, too. They are used to describe real catalyst shapes. The boundary conditions of the differential equation for simultaneous reaction and diffusion depend on the details of the system (see this Chapter and Chapter 6).

The diffusion flux of a component, N_i, is related to the local concentration gradients of the compounds. Depending on the mathematical model used for diffusion, different expressions are obtained.

The pioneering work in diffusion was carried out by Stefan and Maxwell in the nineteenth century. According to the general law of Stefan and Maxwell, the diffusion flux of each component (x_i) is related to the concentration gradients (dc_i/dr):

$$\underline{N} = \underline{F} \frac{d\underline{c}}{dr} \qquad (4.11)$$

where $\underline{F}$ is the coefficient matrix, the structure of which depends on the system. For molecular diffusion, the theory of Stefan and Maxwell gives complicated relationships between the fluxes and concentration gradients (Aris (1975), Fott and Schneider (1984), Salmi and Wärnå (1991)). For porous catalysts, additional effects appear: Knudsen diffusion originated with the molecular collisions with pore walls and surface diffusion. Thus, the vector $\underline{F}$ has additional elements in this case.

A commonly used approach to simplify the mathematical description of diffusion is to introduce some kind of effective diffusion coefficients (D_{ei}), which are formally related to the corresponding concentration gradient only, that is,

$$N_i = -D_{ei} \frac{dc_i}{dr} \qquad (4.12)$$

This concept essentially simplifies the mathematical treatment of diffusion; for instance, for the simplest geometry (the slab or fluid film), we obtain:

$$\varepsilon \frac{dc_i}{dt} = D_{ei} \frac{d^2 c_i}{dr^2} + r_i \alpha \qquad (4.13)$$

However, the validity of the simplification should be investigated case by case.

The concepts of diffusion and reaction will be used as a basis for the treatment of catalytic processes (Chapters 4–5) as well as gas–liquid reactions (Chapter 6).

4.2 Catalytic fixed bed reactors

Catalytic fixed beds are called the work horse of the chemical industry. Most industrially applied chemical reactions are catalytic, and most catalytic reactions are carried out in fixed beds. A fixed bed consists of stagnant catalyst layers organized in such a way that the reacting gas and liquid can easily come into contact with the particles. The basic design of a catalytic fixed bed reactor along with some special constructions is illustrated in Figure 4.3.

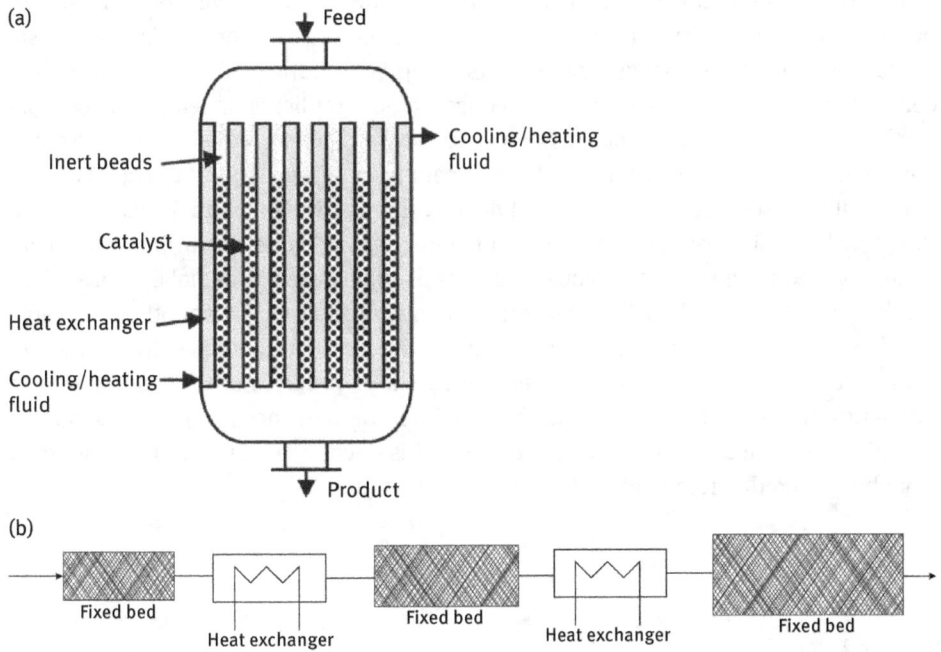

Figure 4.3: Some catalytic fixed bed reactors (Salmi et al. (2011)).

Chemical reactions take place mainly on the internal surfaces, in the pores of catalyst particles. The smaller the particle, the more efficient the process is, since diffusion, resistance in the pores is minimized by decreasing the particle size. The ultimate optimum, a microscopic particle size, is not tractable in fixed beds, however, since this would lead to an infinite pressure drop in the bed. Thus, a practical optimum needs to be found with small enough particles to suppress the diffusion limitation, however still keeping the pressure drop within reasonable limits. A reliable model for a fixed bed reactor thus has to take into account the catalyst particle, the catalyst-fluid interface, the bulk of the fluid phase as well as the interaction of the reactor with its surroundings. In this section, various models for catalytic fixed bed reactors will be considered.

4.2.1 Models for fixed beds

The crucially important issue in the modelling of a fixed bed reactor is to judge whether the diffusion resistance inside the catalyst particles is important or not. If the reaction is slow but diffusion is rapid, no concentration gradients appear in the catalyst particle, the reactor operates in a kinetic regime and it can be described by a so-called pseudo-homogeneous model. Diffusion effects are not visible in a pseudo-homogeneous model, and the reactor can be described with a very similar formalism to that used for homogeneous tube reactors (Chapter 3). If the reaction is rapid but the component diffusion is slow, profound concentration gradients appear in the catalyst particles. This phenomenon has to be described by a reaction-diffusion model for catalyst particles. The model of the particles is coupled to the mass and energy balances of the bulk phases, that is, gas or liquid flowing through the interstitial space between the particles. This description is called a heterogeneous model. A further refinement of homogeneous and heterogeneous models is possible based on radial effects appearing in the catalyst bed. If the reaction is highly exothermic or endothermic, large temperature gradients are found in the axial direction as illustrated in Figure 4.4. Temperature gradients imply that chemical reactions progress at different speeds in different radial locations. This inevitably results in radial concentration gradients. To describe these effects, a two-dimensional model is needed, which comprises not only axial but also radial temperature and concentration gradients.

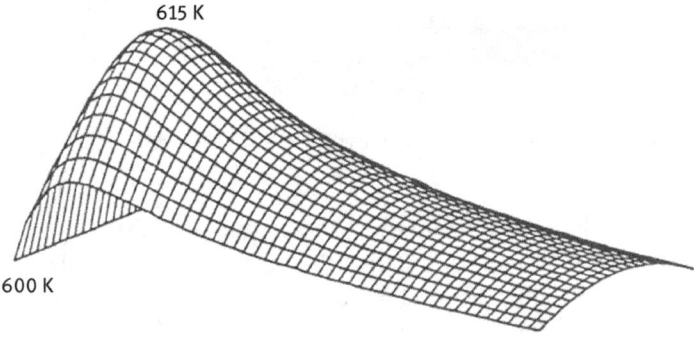

Figure 4.4: Temperature profile in a fixed bed (exothermic reaction with cooling, P. Lundén 1991).

Table 4.2 summarizes the modelling concepts of fixed beds. Below, we will discuss the one and two-dimensional homogeneous models and the one-dimensional heterogeneous model in detail. Based on these concepts, any other model can easily be derived.

Table 4.2: Characterization of models for catalytic fixed bed reactors.

Model	Characteritic features
Pseudo-homogeneous model: diffusion limitations inside the catalyst neglected	
one-dimensional	Plug flow or axial dispersion; neither radial concentration nor temperature gradients in the reactor
two-dimensional	Plug flow or axial dispersion; radial concentration and temperature gradients in the reactor
Heterogeneous model: diffusion resistance in the catalyst notable	
one-dimensional	Plug flow or axial dispersion; neither radial concentration nor temperature gradients in the reactor; concentration and temperature gradients inside the catalyst particles
two-dimensional	Plug flow or axial dispersion; radial concentration and temperature gradients in the reactor; concentration and temperature gradients inside the catalyst particle

4.2.2 Pseudo-homogeneous models for fixed beds

Two and one-dimensional pseudo-homogeneous models are considered in this section. Because the model is pseudo-homogeneous, mass and heat transfer effects inside catalyst particles are ignored. The treatment starts with a two-dimensional model; the one-dimensional model can easily be obtained as a special case of the two-dimensional model by ignoring the radial concentration and temperature gradients.

The reactor volume element used for the derivation of mass and energy balances for the pseudo-homogeneous model is illustrated in Figure 4.5.

The void fraction for the catalyst bed is defined as

$$\varepsilon = \frac{V_F}{V_R} \tag{4.14}$$

where V_F denotes the volume of the fluid in the interstitial volume between the catalyst particles.

Plug flow in axial direction and radial dispersion of heat and mass is presumed. For the sake of simplicity, only steady-state models will be considered below. The basis of the modelling is the cylindrical volume element in the bed (Figure 4.5).

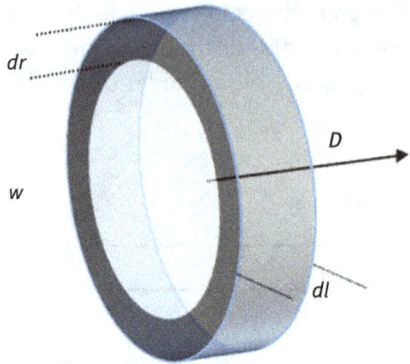

Figure 4.5: Volume element in fixed bed modelling.

The mass balance of a component can be written in the following form,

$$\dot{n}_{i,in} + \left[-\varepsilon 2\pi r \Delta l \frac{d(Dc_i)}{dr}\right]_{in} + \rho_B r_i 2\pi \Delta r \Delta l = \dot{n}_{i,out} + \left[-\varepsilon 2\pi r \Delta l \frac{d(Dc_i)}{dr}\right]_{out} \quad (4.15)$$

where D is the radial dispersion coefficient that presumably has the same value for all components and ρ_B is the bulk density of the catalyst, $\rho_B = m_{cat}/V_R$. We denote differences in molar flows as

$$\Delta \dot{n}_i = \Delta \dot{n}_{i,out} - \Delta \dot{n}_{i,in} = 2\pi r \Delta r \Delta(c_i w) \quad (4.16)$$

Differences between the radial dispersion terms can be expressed through

$$\left[\varepsilon 2\pi r \Delta l \frac{d(Dc_i)}{dr}\right]_{out} - \left[\varepsilon 2\pi r \Delta l \frac{d(Dc_i)}{dr}\right]_{in} = \Delta\left[\varepsilon 2\pi r \Delta l \frac{d(Dc_i)}{dr}\right] \quad (4.17)$$

After inserting relations (4.16) and (4.17) in the original mass balance eq. (4.15), we get:

$$\Delta\left[\varepsilon 2\pi r \Delta l \frac{d(Dc_i)}{dr}\right] + \rho_B r_i 2\pi \Delta r \Delta l = \Delta \dot{n}_i = 2\pi r \Delta r \Delta(c_i w) \quad (4.18)$$

Division of eq. (4.18) by $2\pi \Delta r \Delta l$ gives

$$\frac{\Delta(\varepsilon r d(Dc_i)/dr)}{r \Delta r} + \rho_B r_i = \frac{\Delta(c_i w)}{\Delta l} \quad (4.19)$$

We assume that the dispersion coefficient does not change in the radial direction, $D \approx$ constant and that the void fraction, $\varepsilon \approx$ constant. By allowing $\Delta l \rightarrow 0$ and $\Delta r \rightarrow 0$, the differential equation is obtained,

$$\frac{\varepsilon D}{r}\frac{d}{dr}\left(r\frac{dc_i}{dr}\right)+\rho_B r_i = \frac{d(c_i w)}{dl} \tag{4.20}$$

After carrying out the differentiation of the left-hand side, we obtain

$$\varepsilon D\left(\frac{d^2 c_i}{dr^2}+\frac{1}{r}\frac{dc_i}{dr}\right)+\rho_B r_i = \frac{d(c_i w)}{dl} \tag{4.21}$$

which represents a rather general form of a two-dimensional pseudo-homogeneous model for a fixed bed.

In the solution of the steady-state model, it is conventional to define variable $\dot{n}_i$:

$$\dot{n}_i = c_i w \pi R^2 \tag{4.22}$$

where $\dot{n}_i$ is an artificial molar flow. We denote the volumetric flow rate $\dot{V}=w\pi R^2$, and by recalling the definition

$$c_i = \frac{\dot{n}_i}{\dot{V}} \tag{4.23}$$

and approximating the radial derivatives of the concentrations as follows:

$$\frac{dc_i}{dr} \approx \frac{1}{\dot{V}}\frac{d\dot{n}_i}{dr} \tag{4.24}$$

$$\frac{d^2 c_i}{dr^2} = \frac{1}{\dot{V}}\frac{d^2 \dot{n}_i}{dr^2} \tag{4.25}$$

The steady-state model eq. (4.21) now becomes

$$\frac{d\dot{n}_i}{dl} = \frac{\varepsilon D \pi R^2}{\dot{V}}\left(\frac{d^2 \dot{n}_i}{dr^2}+\frac{1}{r}\frac{d\dot{n}_i}{dr}\right)+\rho_B r_i \pi R^2 \tag{4.26}$$

Dimensionless coordinates are introduced accordingly,

$$l = z\cdot L,\ r = \xi\cdot R \tag{4.27}$$

and the mass balance can be rewritten as

$$\frac{d\dot{n}_i}{dz} = \frac{\varepsilon D V_R}{\dot{V} R^2}\left(\frac{d^2 \dot{n}_i}{d\xi^2}+\frac{1}{\xi}\frac{d\dot{n}_i}{d\xi}\right)+\rho_B V_R r_i \tag{4.28}$$

The benefit of eq. (4.28) is that only one dependent variable ($\dot{n}_i$) exists. The concentrations can always be obtained from eq. (4.23), and the volumetric flow rate is updated from a gas law; for example, the ideal gas flow gives $\dot{V}=\sum \dot{n}_i RT/P$.

The initial and boundary conditions of eq. (4.28) are:

$$\dot{n} = \dot{n}_{i0} \text{ at } z = 0 \tag{4.29}$$

$$\frac{d\dot{n}_i}{d\xi} = 0 \text{ at } \xi = 0 \tag{4.30}$$

$$\frac{d\dot{n}_i}{d\xi} = 0 \text{ at } \xi = 1 \tag{4.31}$$

The boundary condition (4.30) is due to symmetry reasons and the boundary condition (4.31) is due to the fact that no dispersion from the reactor can take place through the reactor wall.

Under plug flow conditions, the radial dispersion coefficient is zero, $D = 0$, and we obtain the simple plug flow model,

$$\frac{d\dot{n}_i}{dz} = \rho_B V_R r_i \tag{4.32}$$

for which the initial condition (4.29) is valid.

It should be observed that the catalyst bulk density is directly related to the mass of the catalyst:

$$\rho_B V_R = m_{cat} \tag{4.33}$$

A numerical solution of eq. (4.28) provides the molar flow (and concentration) profiles. However, we are primarily interested in $\dot{n}_{iout}$, the molar flow at the reactor outlet, which is obtained from

$$\dot{n}_{i,out} = \int_0^R c_i w 2\pi r \, dr \tag{4.34}$$

That is, by numerical integration of the outlet concentration profiles. Equation (4.34) can be expressed as

$$\dot{n}_{i,out} = \int_0^1 \frac{\dot{n}_i}{\pi R^2} 2\pi R^2 \xi \, d\xi \tag{4.35}$$

finally giving the average flow at the outlet:

$$\dot{n}_{i,out} = 2 \int_0^1 \dot{n}_i \xi \, d\xi \tag{4.36}$$

The procedure is straightforward: $\dot{n}_i(\xi)$ is obtained by numerically solving the fundamental balance eq. (4.28) with the initial and boundary conditions (4.29)–(4.31), after which the average molar flow is calculated from eq. (4.36) by numerical integration.

For plug flow conditions, $\dot{n}_i$ is constant in the radial direction and $\dot{n}_{i,out} = \dot{n}_i$. The overall conversion of an arbitrary component (i) is defined as

$$X = \frac{\dot{n}_{i,in} - \dot{n}_{i,out}}{\dot{n}_{i,in}} \tag{4.37}$$

To derive the energy balance, an analogous procedure is applied. The heat conductance in the radial direction is described by the effective heat conductivity (λ) of the bed following the law of Fourier. The axial heat conductance of the bed is ignored.

The steady-state energy balance becomes

$$\left(-\lambda \frac{dT}{dr} 2\pi r \Delta l\right)_{in} + \rho_B \sum_j R_j(-\Delta H_{rj}) 2\pi r \Delta r \Delta l = \left(-\lambda \frac{dT}{dr} 2\pi r \Delta l\right)_{out} + \Delta \dot{m} c_p \Delta T \tag{4.38}$$

The element of the mass flow is

$$\Delta \dot{m} = \rho_0 w_0 2\pi r \Delta r \tag{4.39}$$

and the difference between the heat conductivity terms is denoted by

$$\left(\lambda \frac{dT}{dr} 2\pi r \Delta l\right)_{out} - \left(\lambda \frac{dT}{dr} 2\pi r \Delta l\right)_{in} = \Delta\left(\lambda \frac{dT}{dr} 2\pi r \Delta l\right) \tag{4.40}$$

Division of the energy balance eq. (4.38) by $2\pi r \Delta l \Delta r$ gives

$$\frac{\Delta(\lambda \frac{dT}{dr} r)}{r \Delta r} + \rho_B \sum_j R_j(-\Delta H_{rj}) = c_p \rho_0 w_0 \frac{\Delta T}{\Delta l} \tag{4.41}$$

By assuming constant radial heat conductivity (λ) and allowing $\Delta r \to 0$ $\Delta l \to 0$, the following differential equation is obtained:

$$\frac{\lambda}{r} \frac{d}{dr}\left(r \frac{dT}{dr}\right) + \rho_B \sum_j R_j(-\Delta H_{rj}) = c_p \rho_0 w_0 \frac{dT}{dl} \tag{4.42}$$

After carrying out the differentiation, the energy balance becomes

$$\frac{dT}{dl} = \frac{1}{c_p \rho_0 w_0}\left[\lambda\left(\frac{d^2T}{dr^2} + \frac{1}{r}\frac{dT}{dr}\right) + \rho_B \sum_j R_j(-\Delta H_{rj})\right] \tag{4.43}$$

The use of the dimensionless coordinates z and ξ finally gives

$$\frac{dT}{dz} = \frac{\tau}{c_p \rho_0}\left[\frac{\lambda}{R^2}\left(\frac{d^2T}{d\xi^2} + \frac{1}{\xi}\frac{dT}{d\xi}\right) + \rho_B \sum_j R_j(-\Delta H_{rj})\right] \tag{4.44}$$

where the space time is defined as follows

$$\tau = \frac{L}{w_0} \tag{4.45}$$

The initial condition of the energy balance eq. (4.44) is

$$T = T_0 \text{ at } z = 0 \tag{4.46}$$

Separate boundary conditions are defined for the centreline and the wall of the reactor tube. At the centreline,

$$\frac{dT}{d\xi} = 0 \text{ at } \xi = 0 \tag{4.47}$$

is valid for symmetry reasons.

At the reactor wall ($\xi = 1$), the heat transferred out of the reactor or into the reactor is

$$-\lambda \left(\frac{dT}{dr}\right)_{r=R} 2\pi R \Delta l = U_w 2\pi R \Delta l (T - T_c) \tag{4.48}$$

where U_w is the overall heat transfer coefficient of the reactor wall and T_c is the temperature of the reactor surroundings. After introducing the dimensionless coordinate ($\xi = r/R$), the boundary condition is rewritten as ($dr = R d\xi$)

$$-\frac{\lambda}{R}\left(\frac{dT}{d\xi}\right)_{\xi=1} = U_w(T - T_c) \text{ at } \xi = 1 \tag{4.49}$$

where $R = d_T/2$, d_T is the diameter of the reactor tube.

The reactor model consists of mass and energy balance eqs. (4.28) and (4.44), which are solved together with a correlation for the pressure drop in fixed beds (see Section 4.2.6). The model basically consists of coupled parabolic partial differential equations (PDEs). An example simulation is described in Figure 4.6 with the gas-phase hydrogenation of toluene to methyl cyclohexane.

4.2.3 Heterogeneous model for fixed beds

We consider the porous catalyst particle with simultaneous reaction and diffusion based on the theory presented in Section 4.1. Because of diffusion resistance, concentration and temperature gradients spontaneously appear in the particle (Figure 4.7). A particle with a perfect spherical geometry is considered here, after which a generalization will be made.

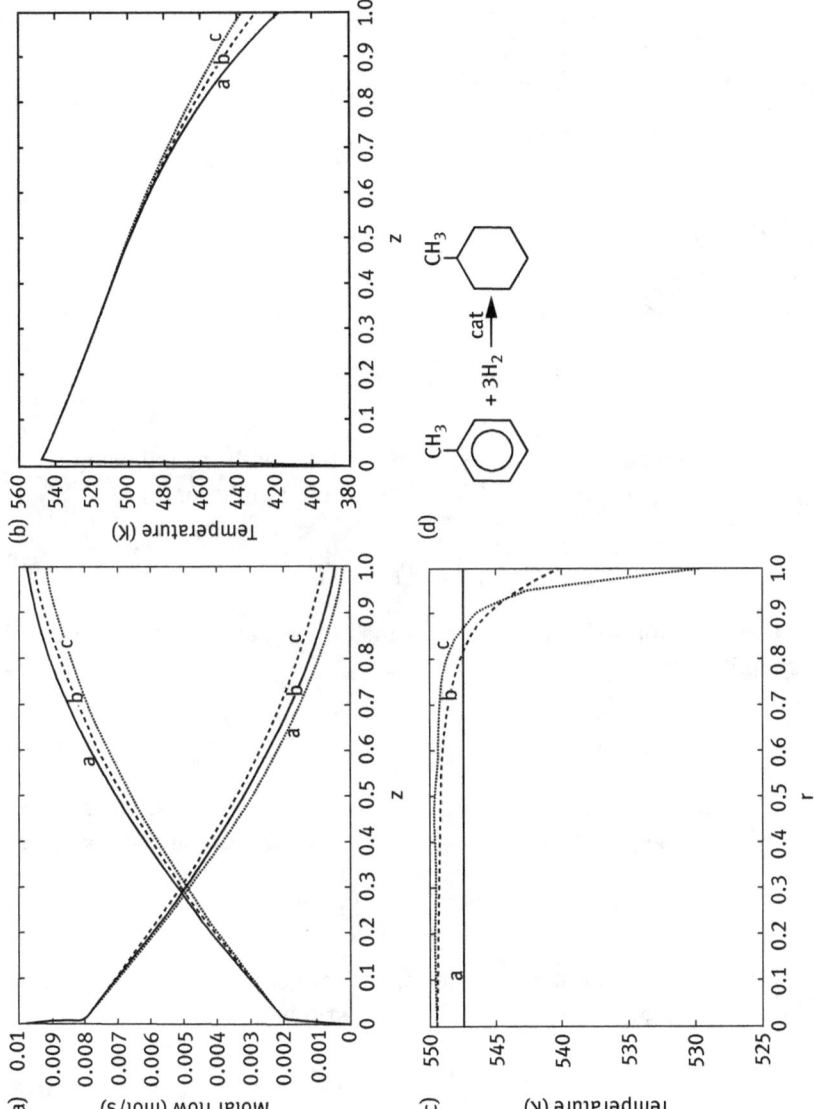

Figure 4.6: Simulation of a two-dimensional model catalytic fixed bed: catalytic hydrogenation of toluene in gas-phase. Molar flows (a), temperatures (b) and radial temperature profiles (c).

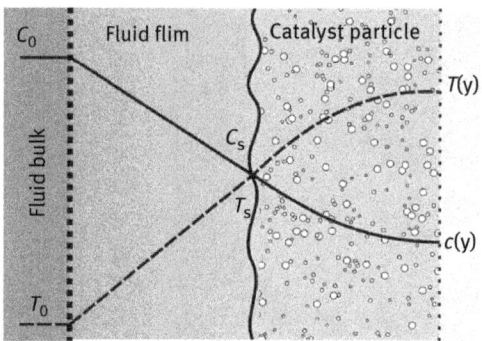

Figure 4.7: Reaction and diffusion in an ideal porous catalyst.

The general mass balance equation, eq (4.10), is used with the following insertions: $\alpha = \rho_p$, $N_i = -D_{ei}(dc_i/dr)$ and $s = 2$. The balance equation becomes

$$\frac{1}{r^2}\frac{d}{dr}\left(D_{ei}r^2\frac{dc_i}{dr}\right) + r_i\rho_p = \varepsilon_p\frac{dc_i}{dt} \qquad (4.50)$$

If D_{ei} can be assumed approximately constant inside the particle, the differential eq. (4.50) is simplified to

$$\frac{d^2c_i}{dr^2} + \frac{2}{r}\frac{dc_i}{dr} + \frac{r_i\rho_p}{D_{ei}} = \frac{\varepsilon_p}{D_{ei}}\frac{dc_i}{dt} \qquad (4.51)$$

Provided that R_c is the characteristic length, we may introduce the dimensionless coordinate $x = r/R_c$. Subsequently, $r = x \cdot R_c$ gives the final form of the mass balance,

$$\frac{d^2c_i}{dx^2} + \frac{2}{x}\frac{dc_i}{dx} + \frac{r_i\rho_p R_c^2}{D_{ei}} = \frac{\varepsilon_p R_c^2}{D_{ei}}\frac{dc_i}{dt} \qquad (4.52)$$

The equations presented above are valid for spherical geometry. It can easily be shown that for a general geometry, eq. (4.52) can be written in the form

$$\frac{1}{r^{a-1}}\frac{d}{dr}\left(D_{ei}r^{a-1}\frac{dc_i}{dr}\right) + r_i\rho_p = \varepsilon_p\frac{dc_i}{dt} \qquad (4.53)$$

which in a straightforward way gives, for the steady state:

$$\frac{d^2c_i}{dx^2} + \frac{(a-1)}{x}\frac{dc_i}{dx} + \frac{r_i\rho_p R_c^2}{D_{ei}} = 0 \qquad (4.54)$$

where $a = s+1$ and s are the shape factors of the catalyst; a is generally defined by $a = (A_P/V_P) \cdot R_C$, where A_p is the outer surface area of the catalyst and V_p is the volume of the catalyst particle. The values of the shape factors are listed in Table 4.3 and illustrated in Figure 4.8.

Table 4.3: Shape factors for various particle diameters.

Geometry	R_c	a	s
Sphere	radius	3	2
Long cylinder	radius	2	1
Cylinder with $h = 2R$	radius	3	2
Cylinder with $h = nR, n > 1$	radius	$2(n+1)/n$	$a - 1$
Slab	half thickness	1	0

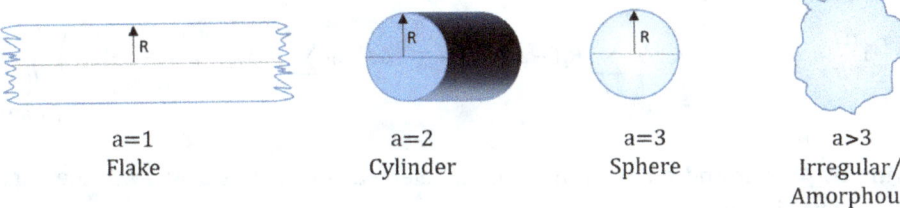

| a=1 | a=2 | a=3 | a>3 |
| Flake | Cylinder | Sphere | Irregular/Amorphous |

Figure 4.8: Ideal and non-deal particle geometries.

For non-ideal geometries, a non-integer value of the shape factor is selected. The diffusion term can be written in a general form by using the Nabla operator (∇):

$$\frac{1}{r^s} \frac{d}{dr}\left(r^s \frac{dc_i}{dr}\right) = \nabla^2 c_i \tag{4.55}$$

In a steady state, the concentrations in eq. (4.54) are constant with respect to time:

$$\frac{dc_i}{dt} = 0 \tag{4.56}$$

The energy balance for an isotropic and spherical catalyst particle can be obtained in a completely analogous manner. Effective heat conductivity is assumed. Consequently, the energy balance for a spherical element is obtained as

$$\left(-\lambda \frac{dT}{dr} 4\pi r^2\right)_{in} + \rho_p 4\pi r^2 \Delta r \sum_j R_j (-\Delta H_{rj}) = \left(-\lambda \frac{dT}{dr} 4\pi r^2\right)_{out} + \frac{dU}{dt} \tag{4.57}$$

Provided that the temperatures of the fluid and the solid material are equal, the accumulation of internal energy (dU/dt) can be expressed by

$$\frac{dU}{dt} = \left(\sum_i n_i c_{vmi}\right) \frac{dT}{dt} + c_{vs} m_s \frac{dT}{dt} \tag{4.58}$$

where c_{vs} is the heat capacity of the solid material, and m_s is the mass of the solid material. For the amount of substance (n_i) and the mass of the catalyst (m_s), we have:

$$n_i = c_i \varepsilon_P 4\pi r^2 \Delta r \tag{4.59}$$

$$m_s = \rho_s(1-\varepsilon_P)4\pi r^2 \Delta r \tag{4.60}$$

where ρ_s denotes the skeletal density of the solid material. The difference between heat conduction terms is denoted by

$$\left(\lambda \frac{dT}{dr} 4\pi r^2\right)_{out} - \left(\lambda \frac{dT}{dr} 4\pi r^2\right)_{in} = \Delta\left(\lambda \frac{dT}{dr} 4\pi r^2\right) \tag{4.61}$$

and the energy balance (4.57) becomes

$$\Delta\left(\lambda \frac{dT}{dr} 4\pi r^2\right) + \rho_P 4\pi r^2 \Delta r \sum_j R_j(-\Delta H_{rj}) = 4\pi r^2 \Delta r \left(\varepsilon_P \sum_i c_i c_{vmi} + \rho_s(1-\varepsilon_P)c_{vs}\right)\frac{dT}{dt} \tag{4.62}$$

Diving by $4\pi r^2 \Delta r$ and allowing the volume element to diminish, $\Delta r \to 0$, the very general form is obtained

$$\frac{1}{r^2}\frac{d}{dr}\left(\lambda r^2 \frac{dT}{dr}\right) + \rho_P \sum_j R_j(-\Delta H_{rj}) = \left(\varepsilon_P \sum_i c_i c_{vmi} + \rho_s(1-\varepsilon_P)c_{vs}\right)\frac{dT}{dt} \tag{4.63}$$

Assuming constant heat conductivity (λ) and carrying out the differentiation leads to

$$\lambda\left(\frac{d^2 T}{dr^2} + \frac{1}{r}\frac{dT}{dr}\right) + \rho_P \sum_j R_j(-\Delta H_{rj}) = \left(\varepsilon_P \sum_i c_i c_{vmi} + \rho_s(1-\varepsilon_P)c_{vs}\right)\frac{dT}{dt} \tag{4.64}$$

The use of the dimensionless coordinate finally gives

$$\frac{d^2 T}{dx^2} + \frac{2}{x}\frac{dT}{dx} + \frac{\rho_P R_c^2}{\lambda}\sum_j R_j(-\Delta H_{rj}) = \frac{R_c^2}{\lambda}\left(\varepsilon_P \sum_i c_i c_{vmi} + \rho_s(1-\varepsilon_P)c_{vs}\right)\frac{dT}{dt} \tag{4.65}$$

For an arbitrary geometry, the energy balance can be generalized as

$$\frac{d^2 T}{dx^2} + \frac{(a-1)}{x}\frac{dT}{dx} + \frac{\rho_P R_c^2}{\lambda}\sum_j R_j(-\Delta H_{rj}) = \frac{R_c^2}{\lambda}\left(\varepsilon_P \sum_i c_i c_{mi} + \rho_s(1-\varepsilon_P)c_{vs}\right)\frac{dT}{dt} \tag{4.66}$$

where a is explained in Table 4.3.

The accumulation term $\frac{R_c^2}{\lambda}\left(\varepsilon_P \sum_i c_i c_{vmi} + \rho_s(1-\varepsilon_P)c_{vs}\right)\frac{dT}{dt}$ in the above equations is a rather complex one, involving several physical parameters, but the heat capacity in the accumulation term is often approximated by

$$\varepsilon_P \sum_i c_i c_{vmi} + (1-\varepsilon_P)\rho_s c_{vs} \approx c_p \rho_P \tag{4.67}$$

giving for an arbitrary geometry

$$\frac{c_p \rho_p R_c^2}{\lambda} \frac{dT}{dt} = \frac{d^2T}{dx^2} + \frac{(a-1)}{x}\frac{dT}{dx} + \frac{\rho_p R_c^2}{\lambda}\sum_j R_j(-\Delta H_{rj}) \tag{4.68}$$

In steady-state conditions, the time derivative vanishes, that is, dT/dt, and we obtain

$$\frac{d^2T}{dx^2} + \frac{(a-1)}{x}\frac{dT}{dx} + \frac{\rho_p R_c^2}{\lambda}\sum_j R_j(-\Delta H_{rj}) = 0 \tag{4.69}$$

In a mathematical sense, the energy balance equation is completely analogous with the mass balance equation for a porous layer, eq. (4.54).

Below, we will consider the boundary conditions of the mass and energy balances. On the outer surface of the particle, mass transfer through the fluid film and the particle is equal (Figure 4.9), giving the important equation,

$$k_{Gi}(c_{0i} - c_i)4\pi R_c^2 = D_{ei}\left(\frac{dc_i}{dr}\right)_{r=R_c} 4\pi R_c^2 \tag{4.70}$$

which leads to

$$\frac{dc_i}{dr} = \frac{k_{Gi}(c_{0i} - c_i)}{D_{ei}} \text{ at } r = R_c \tag{4.71}$$

where k_{Gi} denotes the mass transfer coefficient for the fluid phase, and c_{0i} is the concentration of the component (i) in the bulk phase. The introduction of the dimensionless coordinate gives

$$\frac{dc_i}{dx} = \frac{k_{Gi}R_c}{D_{ei}}(c_{0i} - c_i) \text{ at } x = 1 \tag{4.72}$$

In the centre of the particle, the boundary condition is

$$\frac{dc_i}{dx} = 0 \text{ at } x = 0 \tag{4.73}$$

for symmetry reasons.

For heat transfer, we analogously have on the outer surface of the particle the boundary condition

$$h(T - T_0)4\pi R_c^2 = -\lambda\left(\frac{dT}{dr}\right)4\pi R_c^2 \tag{4.74}$$

where h denotes the heat transfer coefficient of the fluid film surrounding the pellet and T_0 is the bulk phase temperature (Figure 4.9).

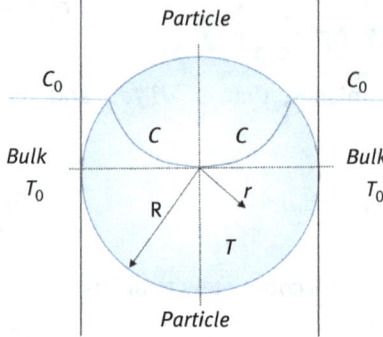

Figure 4.9: Illustration of the boundary conditions for a porous particle.

Equation (4.74) gives

$$\frac{dT}{dx} = \frac{hR_c}{\lambda}(T_0 - T) \text{ at } x = 1 \qquad (4.75)$$

In the centre of the particle, the boundary condition again is because of symmetry reasons:

$$\frac{dT}{dx} = 0 \text{ at } x = 0 \qquad (4.76)$$

4.2.3.1 Special case

For example, for linear kinetics, that is, zero-order $r_i = v_i k$ and first-order $r_i = v_i k c_i$ rate equations, under isothermal conditions, it is possible to solve the steady-state form of the second-order differential eq. (4.54) analytically for slab and spherical geometries. The solutions, that is, the concentration profiles, are illustrated in Figure 4.8. For details of the treatment, as well as semi-analytical solutions of the diffusion equation, the reader is referred to the monumental work of Aris (1975).

Analytical solutions of concentration profiles (Figure 4.10) for first-order kinetics in porous particles are presented below.

$$s = 0 \text{ (slab)} \quad c_i = c_i^s \frac{\cosh(\phi x)}{\cosh(\phi)} \qquad (4.77)$$

$$s = 2 \text{ (sphere)} \quad c_i = c_i^s \frac{\sinh(\phi x)}{x \cdot \sinh(\phi)} \qquad (4.78)$$

$$\phi = R_c \sqrt{\frac{k \cdot \rho_P}{D_{ei}}} \qquad (4.79)$$

where ϕ is the famous Thiele modulus.

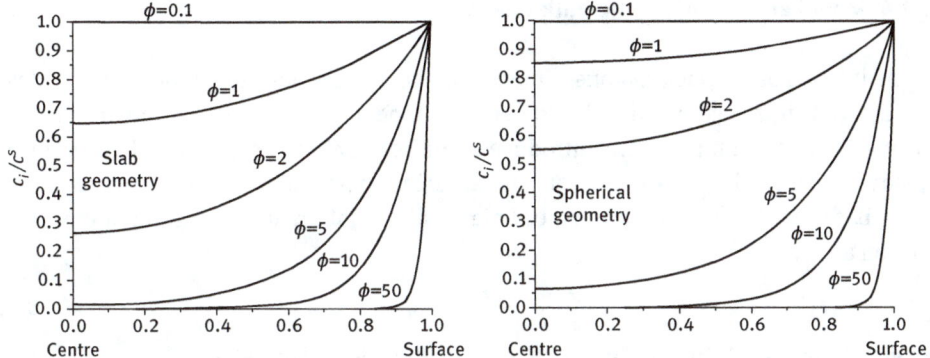

Figure 4.10: Concentration profiles inside a porous catalyst for first order reactions.

Two important dimensionless quantities appeared spontaneously in the derivation of the mass and energy balances, namely,

$$Bi_M = \frac{k_{Gi}R_c}{D_{ei}}, \quad \text{Biot number of mass transfer} \quad (4.80)$$

and

$$Bi = \frac{hR_c}{\lambda}, \quad \text{Biot number of heat transfer} \quad (4.81)$$

These numbers have a clear physical significance: they tell us to what extent the film (the nominator) and the particle itself (the denominator) contribute to mass and heat transfer resistances. Typically, the value of Bi_M is high (> 100), emphasizing the role of mass transfer resistance inside the particle, not in the film. The opposite is true for Bi: heat conductivity of the particle (λ) is high compared to the product of the heat transfer coefficient (h) of the film and the particle radius (R_c). Thus, the main heat transfer resistance is typically located in the film surrounding the catalyst particle. The situation is illustrated in Figure 4.11.

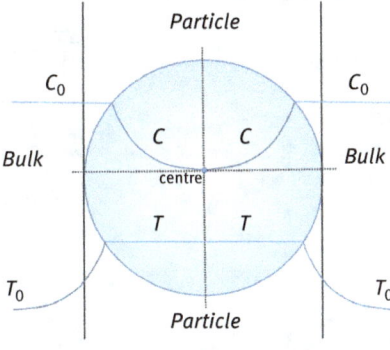

Figure 4.11: Mass and heat transfer resistances in catalyst particles.

4.2.4 Model equations for the bulk phase

In the interest of simplicity, a one-dimensional heterogeneous model for a fixed bed is discussed in this section. In order to keep the model simple and illustrative, the following basic assumptions are introduced: the reactor tube is presumed to be in a steady-state and plug flow conditions are assumed to prevail for the bulk phase of the fluid. Based on these assumptions, the mass balance of an arbitrary component (i) is written as

$$\dot{n}_{i,in} = N_i \Delta A + \dot{n}_{i,out} \qquad (4.82)$$

where N_i denotes diffusion flux at $r = R_c$. The flux, N_i, is defined as

$$N_i = D_{ei} \left(\frac{dc_i}{dr} \right)_{r=R_c} \text{ at } r = R_c \qquad (4.83)$$

As usual, we define the difference between outlet and inlet molar flows:

$$\dot{n}_{i,out} - \dot{n}_{i,in} = \Delta \dot{n}_i \qquad (4.84)$$

The balance thus becomes

$$\Delta \dot{n}_i = -N_i \Delta A \qquad (4.85)$$

Furthermore, the ratio between the mass/heat transfer area and the reactor volume is defined as

$$a_p = \frac{\Delta A}{\Delta V} \qquad (4.86)$$

that is, the mass transfer area/reactor volume is denoted as a_0.

Allowing the volume element to shrink, $\Delta V \to 0$, the mass balance becomes

$$\frac{d\dot{n}_i}{dV} = -N_i a_p \qquad (4.87)$$

The introduction of a dimensionless coordinate, $z = V/V_R$ gives

$$\frac{d\dot{n}_i}{dz} = -N_i a_p V_R \qquad (4.88)$$

where the interfacial flux is given by

$$N_i = \frac{D_{ei}}{R_c} \left(\frac{dc_i}{dx} \right)_{x=1} \qquad (4.89)$$

The initial condition of the bulk phase balance is

$$\dot{n}_i = \dot{n}_{0i} \text{ at } z = 0 \qquad (4.90)$$

The energy balance is obtained in an analogous way by considering the volume element again as illustrated in Figure 4.12.

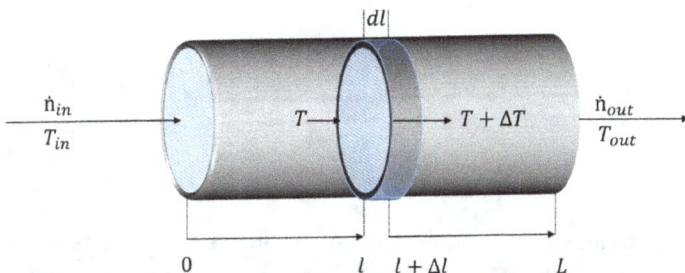

Figure 4.12: Volume element for one-dimensional heterogeneous model.

The heat flux (M) is defined by

$$M = -\lambda \left(\frac{dT}{dr}\right)_{r=R_c} \qquad (4.91)$$

The energy balance becomes

$$M \Delta A = \left(\sum_i c_{pmi} \dot{n}_i\right) \Delta T + \Delta \dot{Q} \qquad (4.92)$$

We introduce $\Delta A = a_0 \Delta V$ and allow $\Delta V \to 0$. The differential equation

$$\frac{dT}{dV} = \frac{1}{\sum_i c_{pmi} \dot{n}_i} \left(M a_0 - \frac{d\dot{Q}}{dV}\right) \qquad (4.93)$$

is obtained. For the heat transfer term $(d\dot{Q}/dV)$, a typically applied expression is

$$\frac{d\dot{Q}}{dV} = U(T - T_c)\frac{dS}{dV} \qquad (4.94)$$

where U is the overall heat transfer coefficient and T_c denotes the temperature of the surroundings.

For tubular reactors with a constant diameter, the following relationship is valid:

$$\frac{dS}{dV} \approx \frac{S}{V_R} = a = \frac{4}{d_T} \qquad (4.95)$$

where d_T is the diameter of the reactor tube.

After inserting the dimensionless coordinate, $z = V/V_R$, the energy balance becomes

$$\frac{dT}{dz} = \frac{V_R}{\sum_i c_{pmi} \dot{n}_i}(Ma_0 - U(T - T_c)a) \qquad (4.96)$$

where

$$M = -\frac{\lambda}{R_c}\frac{dT}{dx} \text{ at } x = 1 \qquad (4.97)$$

The initial condition is

$$T = T_0 \text{ at } z = 0 \qquad (4.98)$$

The one-dimensional, heterogeneous, steady-state model for fixed beds consists of the mass and energy balances (4.88–4.96) for the bulk phase, and the corresponding balances (4.54) and (4.69) for the catalyst particles. The mutual coupling of the balances is discussed in detail in the next section Coupling of the particle and the bulk-phase balance equations.

The derivatives, $(dT/dx)_{x=1}$ and $(dc_i/dx)_{x=1}$, on the outer surface of the particle are obtained by solving the mass and energy balances of the catalyst particle. A special technique that is useful in obtaining an accurate numerical solution will be introduced in this section.

The mass balance of an arbitrary component in the particle can be written as

$$\frac{1}{x^{a-1}}\frac{d}{dx}\left(\frac{dc_i}{dx}\right) + \frac{\rho_p R_c^2}{D_{ei}}r_i = 0 \qquad (4.99)$$

which can be formally integrated,

$$\int_0^y d\left(\frac{dc_i}{dx}\right) = -\frac{\rho_p R_c^2}{D_{ei}}\int_0^1 r_i x^{a-1} dx \qquad (4.100)$$

At $x = 0$, $dc_i/dx = 0$, we obtain

$$y = -\frac{\rho_p R_c^2}{D_{ei}}\int_0^1 r_i x^{a-1} dx \qquad (4.101)$$

where the variable y de facto is

$$y = \left(\frac{dc_i}{dx}\right)_{x=1} \qquad (4.102)$$

The concentration gradients at location $x=1$ are thus obtained from the remarkably simple equation,

$$\left(\frac{dc_i}{dx}\right)_{x=1} = -\frac{\rho_p R_c^2}{D_{ei}} \int_0^1 r_i x^{a-1} dx \tag{4.103}$$

that is, from an integrated average of the generation rate. The flux which is defined as

$$N_i = \frac{D_{ei}}{R_c} \left(\frac{dc_i}{dx}\right)_{x=1} \tag{4.104}$$

is now calculated from

$$N_i(x=1) = -\rho_p R_c \int_0^1 r_i x^{a-1} dx \tag{4.105}$$

An analogous approach is applied to the energy balance:

$$\frac{1}{x^{a-1}} \frac{d}{dx}\left(\frac{dT}{dx}\right) + \frac{\rho_p R_c^2}{\lambda} \sum_j R_j(-\Delta H_{rj}) = 0 \tag{4.106}$$

Integration gives the temperature gradient in a straightforward way,

$$\left(\frac{dT}{dx}\right)_{x=1} = -\frac{\rho_p R_c^2}{\lambda} \int_0^1 \sum_j R_j(-\Delta H_{rj}) x^{a-1} dx = 0 \tag{4.107}$$

The heat flux thus becomes

$$M(x=1) = \rho_p R_c \int_0^1 \sum_j R_j(-\Delta H_{rj}) x^{a-1} dx \tag{4.108}$$

The derivations presented above demonstrate that the molar and heat fluxes on the surface of the catalyst particle can be obtained as integrated averages of the generation rates and products of reaction rates and reaction enthalpies, respectively. The approach is not only interesting from a theoretical point of view but extremely useful for numerical computations: the evaluation of concentration and temperature gradients is based on the computation of integrals involving information from the entire particle. Integration of the concentration and temperature profiles suppresses errors in the numerical solution of balance equations, whereas a numerical differentiation of the solutions (dc_i/dr, dT/dr) accentuates them. Therefore, the use of eqs (4.105) and (4.108) is strongly preferred. The fluxes are obtained with a higher precision from the integrals than from the derivatives.

4.2.5 Pressure drop in fixed beds

Pressure drop plays an important role in all kinds of fixed beds, even in fixed beds used on the laboratory scale for obtaining kinetic data. For large-scale fixed reactors, the catalyst particle sizes can reach values of up to 1 cm or more. In laboratory-scale fixed beds, considerably smaller catalyst particles are used (even $\ll 1$ mm), which means that – even though laboratory reactors are small – the pressure drop might be considerable, since the particles are small. A general form of the pressure drop equation can be written as

$$\frac{dP}{dl} = -f\frac{\rho w^2}{\varphi' d_p} \qquad (4.109)$$

where d_p is the particle diameter and φ' accounts for the sphericity of the particle ($\varphi' = 1$ for perfect spheres). The factor ρw is constant throughout the reactor ($\rho w = \dot{m}/A = G$) and the pressure gradient thus is de facto proportional to the superficial velocity in the first power (w). Various expressions have been proposed in literature for the friction factor (f). A few of them are collected below. The pressure drop equation is coupled to the mass and energy balances and can be solved numerically together with them.

$$f = \frac{(1-\varepsilon)^2}{\varepsilon^3}\frac{150}{Re} \quad \text{Ergun} \qquad (4.110)$$

$$f = 1.75 \quad \text{Burke and Plummer} \qquad (4.111)$$

$$f = \frac{(1-\varepsilon)^2}{\varepsilon^3}\cdot\frac{a}{(\varphi'\cdot Re)} + \frac{(1-\varepsilon)b}{\varepsilon^3} \quad \text{Ergun, improved}$$

$$a = 150, \ b = 1.75, \ Re = d_p G/\mu \qquad (4.112)$$

4.3 Numerical solution of fixed bed models

In this Chapter, numerical solution strategies and suitable algorithms are reviewed. Pseudo-homogeneous models and heterogeneous models are very different in their character; therefore they are treated separately below.

4.3.1 Solution of pseudo-homogeneous models

The pseudo-homogeneous model consists of PDEs which describe the mass and energy balances and ordinary differential equations (ODEs) describing the pressured drop in the bed. The bottom line is that the pseudo-homogeneous two-dimensional model is an extension – or a correction – of the pseudo-homogeneous one-dimensional model, the longitudinal effects being in any case the dominant ones. Very good numerical methods exist for solving ODEs, initial value problems (IVP), as discussed in connection with homogeneous reactors (Chapter 3).

Thus, a natural approach to the coupled PDE-ODE system describing the fixed bed reactor is to transform the PDEs to ODEs by discretization of the radial coordinate. The derivatives dc/dr, dT/dr, d^2c/dr^2, dT^2/dr^2 can be approximated either with finite differences or by approximation functions, and an extensive system of ODEs is the result.

The finite difference method was introduced in Chapter 3, where useful finite differences relations were discussed in section 3.5.1. It is important to emphasize the fact that central difference formulae should be used to approximate the first and second derivatives. For instance, the simplest central difference formulae for first and second derivatives are listed below:

$$\frac{dy_i}{dx} = \frac{1}{2\Delta x}(y_{i+1} - y_{i-1}) \tag{4.113}$$

$$\frac{d^2 y_i}{dx^2} = \frac{1}{(\Delta x)^2}(y_{i-1} - 2y_i + y_{i+1}) \tag{4.114}$$

The accuracy of the calculation can be improved by using higher order central difference formulae (Table 6.7).

An alternative to finite differences is to use an approximation function. A particularly simple and popular approximation function is a polynomial. This method is called collocation. The accuracy of the approximation can in principle be improved by using higher degree polynomials. However, the risk for oscillations increases as the degree of the polynomial is increased. This is the case when equidistant approximation polynomials are used. Therefore, a special form of collocation has been developed. The requirement is that the polynomial fulfils the differential equation exactly at certain, non-equidistant collocation points (collocation abscissae). The collocation points are the zeros of orthogonal polynomials, typically Legendre and Laguerre polynomials.

This method is called orthogonal collocation, and it was developed into a useful form for chemical engineering particularly by the Wisconsin (W. Stewart) and Lyngby (J. Villadsen and M.L. Michelsen) schools. The approximation polynomial is described in the form of a Legendre polynomial, which implies that the collocation ordinates (the y-values, e.g. concentrations or temperatures) appear in the polynomial, but not the "normal" coefficients (such as $f(x) = a_0 + a_1 x + a_2 x^2$). This is a much more feasible approach, because one has some idea of the order of magnitude of the y values, based on physical and chemical knowledge of the system. The values of the first and second derivatives can be calculated with automatic routines originally published by Villadsen and Michelsen (1978).

4.3.2 Solution strategy of heterogeneous models

The heterogeneous model for a catalytic fixed bed consists of two kinds of equations: the balance equations for the catalyst pellet and the balances for the bulk phases of gas or liquid. The former ones give a boundary value problem (BVP), while the latter one results in an initial value problem (IVP). For the bulk phase, the mass balances are of the type $d\langle int\_u\rangle y\langle /int\_u\rangle/dz = f(\langle int\_u\rangle y\langle /int\_u\rangle)$, where $\underline{y} = [\underline{C}, T]$ and they can be integrated forwards with the aid of the robust algorithms for stiff ODEs (Appendix A.1). However, the fluxes of the components (N_i) from the catalyst particles are included in the bulk-phase balances. They are obtained from a numerical solution of the particle equations, eq. (4.54). Thus, a robust algorithm alternates between the particle balances and the bulk-phase balances. The component fluxes which are needed for the bulk phase balances are updated with the pellet balances. The algorithm is sketched in Figure 4.13.

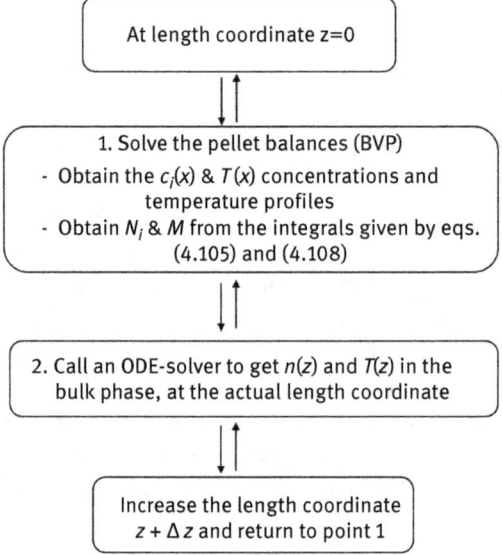

Figure 4.13: Solution of a heterogeneous fixed bed model.

For the solution of the particle balances, spline collocation (orthogonal collocation on finite elements) might be the most powerful method, while the solution of the bulk phase balances is efficiently carried out by the backward difference method suitable for stiff differential equations (Appendix A.2).

The methodology is illustrated with a case study, production of methanol through a gas-phase reaction between carbon monoxide and hydrogen. Besides the

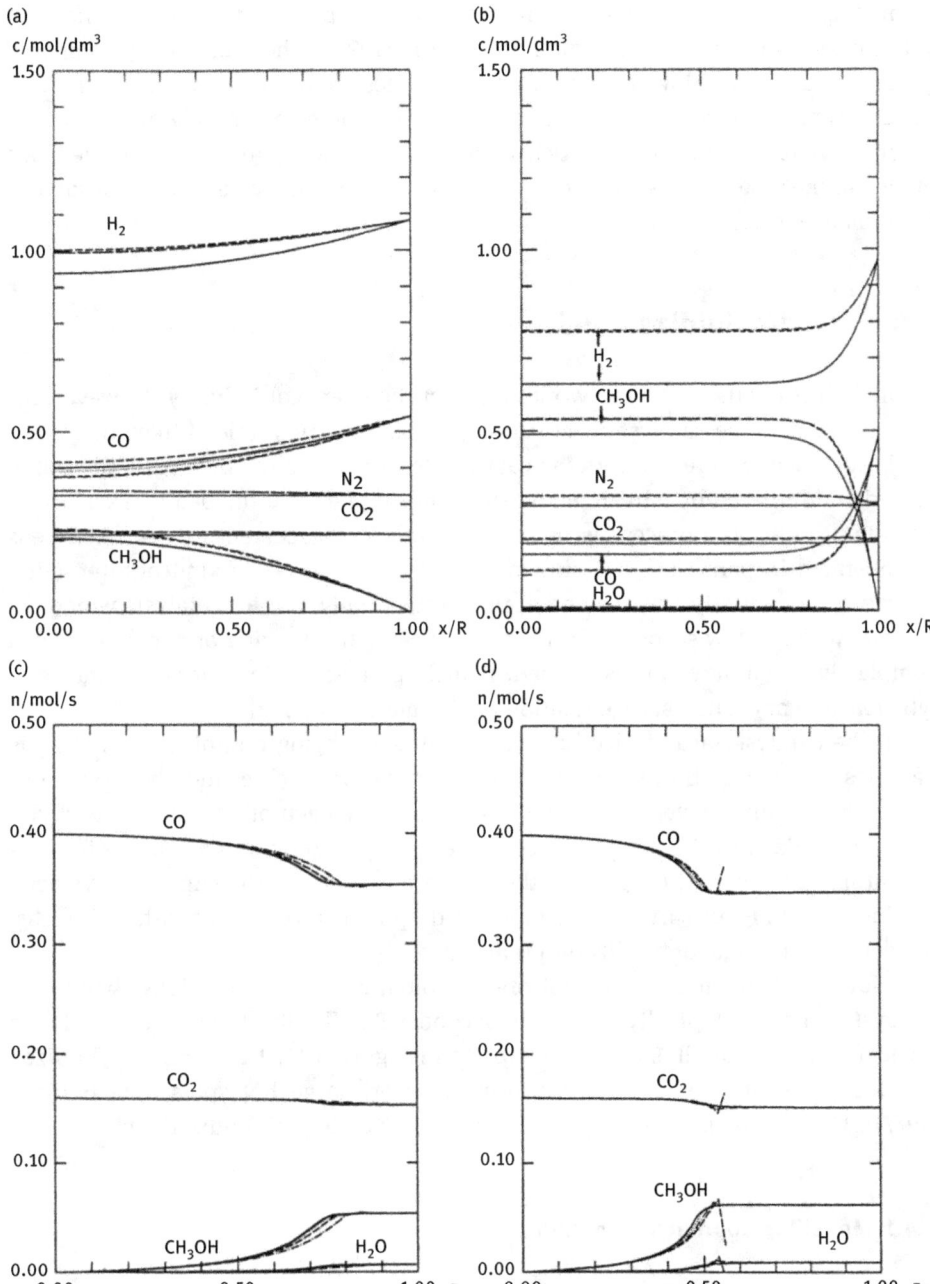

Figure 4.14: Heterogeneous one-dimensional model: methanol synthesis in a fixed bed (Salmi and Wärnå (1991)). Concentration profiles in the catalyst particle (a,b) and molar amounts (c,d).

main reaction, a reverse water-gas shift reaction takes place in the system, since CO_2 is always present in the feed (Salmi and Wärnå (1991)). The numerically computed concentration (molar flow) and temperature profiles are presented in Figure 4.14. As the computed concentration profiles inside the particle reveal, strong diffusion limitation prevails in the catalyst particles. Thus, it is necessary to use a heterogeneous model for the system. A pseudo-homogeneous model would overestimate the catalyst performance by far.

4.4 Catalytic fluidized beds

Catalytic fixed beds act as the work horses of the chemical industry. However, for processes where the catalyst deactivation plays a prominent role, a fixed bed is not the best choice for a reactor. As the catalyst deactivates in the bed, it should be removed and replaced by a fresh, active one. This implies that the bed is taken out of operation. Either the process is halted for catalyst replacement or two fixed beds are operated in parallel. For a slow deactivation process, this kind of alternating operating mode is still feasible and applied industrially, but for catalysts which deactivate within a few seconds, for instance in catalytic cracking of hydrocarbons, a completely new innovation is needed, namely process equipment where the catalytic reaction and catalyst regeneration are intimately coupled.

If the catalyst particle size in a fixed bed keeps being diminished, finally the particles become mobile in the bed. The gravitational force and the drag force caused by the upflow velocity of the gas compensate each other, and the particles become fluidized in the bed. This can be recorded visually, but more precisely by measuring the pressure drop along the bed as a function of the superficial velocity as illustrated in Figure 4.15. As the pressure drop remains constant with an increasing velocity, the state of fluidization is attained.

Fluidized beds made their first breakthrough in catalytic cracking, where the catalyst lifetime is typically limited to seconds. The fluidized bed is coupled to a regeneration unit as illustrated principally in Figure 4.16. Later on, the fluidized bed technology has evolved into processes where large hot spots appear (Rase (1977)). This section is devoted to mathematical modelling of fluidized beds.

4.4.1 Modelling approaches to fluidized beds

For fluidized beds, several mathematical models have been proposed in the past: ideal flow models such as backmixing and plug flow models, axial dispersion models as well as models based on residence time distributions. Generally speaking, these attempts have not been very successful, since they do not account for the real hydrodynamics of the system and the uneven catalyst distribution in fluidized beds. The

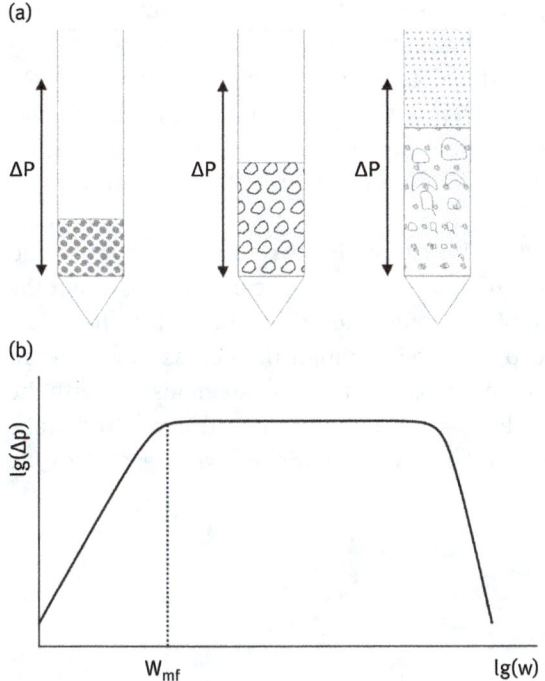

Figure 4.15: Pressure drop in a fixed bed – the state of fluidization.

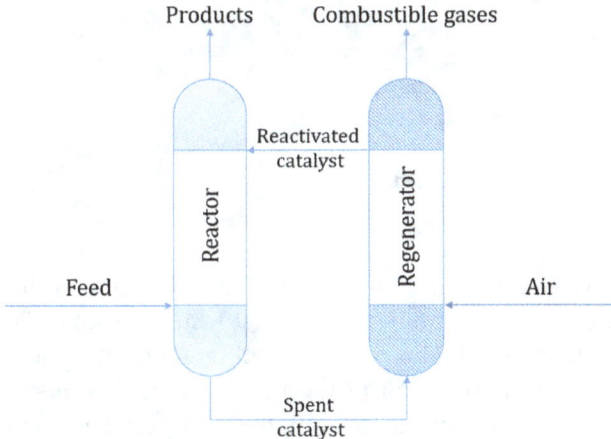

Figure 4.16: Principal construction of a fluidized bed reactor-catalyst regenerator system (Salmi et al. (2011)).

effect of fluidization is easily recognized by measuring the pressure drop over the catalyst bed: as long as the catalyst particles remain stagnant, the pressure drop is proportional to the square of one superficial velocity, as suggested by eq. (4.109). At the moment of fluidization, the catalyst particles become mobile, and the increase in superficial velocity no longer increases the pressure drop (Figure 4.15). The superficial velocity and the bed void fraction valid under those conditions are called w_f and ε_p.

Intuitively, one might assume that the gas and the catalyst particles would form a pseudo-homogeneous dispersion on a macroscopic scale, but this is not the case. In reality, a fluidized bed resembles a boiling liquid: bubbles which have less of catalyst particles are formed and transported through the bed as illustrated in Figure 4.17. The bulk phase prevailing in the reactor is called an emulsion. With the bubbles, some catalyst is transported; these sections are called cloud and wake phases (Figure 4.17). Particularly the wake phase is enriched with respect to the catalyst.

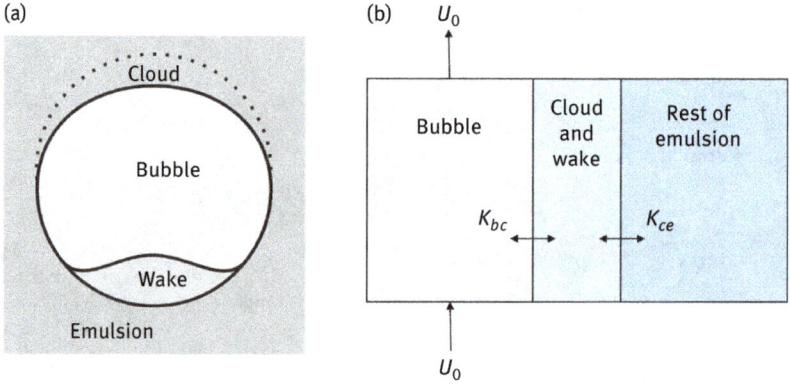

Figure 4.17: Hydrodynamics and phases of a fluidized bed.

The discussion presented above clearly reveals that a realistic description of fluidized beds should be based on a true hydrodynamic model which accounts for the existence of different regions in the fluidized bed with different catalyst bulk densities as well as an appropriate description of the interfacial mass transfer between the regions. A pioneering effort in the modelling of fluidized beds was carried out by Kunii and Levenspiel, who were the first to propose a realistic hydrodynamic model for fluidized beds. After their pioneering publication, several extensions of the model have been proposed: a summary is provided by Levenspiel (1999). Here we stay with the basic features of the model – extensions can easily be made and they are not essential for the understanding of the fundamental principles.

4.4.2 Kunii-Levenspiel model of fluidized beds

In this section, the basic principles of the Kunii-Levenspiel model for fluidized beds are introduced. Different variations of the original model have been published, but here we will keep to a simple and illustrative one. The basic assumptions of the Kunii-Levenspiel model are shortly stated below.

1) The bubbles are assumed to have a uniform size distribution, that is, only one bubble size is considered. This assumption is not of crucial importance, since a bubble size distribution can be easily implemented in the model.
2) Each bubble has a cloud and wake phase with it, moving together with the bubble.
3) For the emulsion phase, minimum fluidization conditions prevail.
4) Chemical reactions proceed in all phases (bubbles, cloud & wake, emulsion), but the catalyst bulk densities of the phases are different.
5) Clouds and wakes are considered as a single phase in the simplified version of the Kunii-Levenspiel model.
6) The mass transfer rates between the phases are directly proportional to the concentration differences.

Based on the hypotheses presented above, the mass balances for an arbitrary component in different phases will be derived in the following sections.

4.4.2.1 Bubble phase

The Kunii-Levenspiel model is illustrated in Figure 4.17. The mass balances of each phase will be derived below. A general volume element in the bed is defined as

$$\Delta V = \Delta V_b + \Delta V_c + \Delta V_e \tag{4.115}$$

where b, c and e denote bubble, cloud & wake and emulsion phases, respectively.

For the bubble phase, the mass balance of component (i) is

$$\dot{n}_{bi,in} + r_{bi}\rho_{Bb}\Delta V_b = \dot{n}_{bi,out} + K_{bci}(c_{bi} - c_{ci})\Delta V_b \tag{4.116}$$

Equation (4.116) states that a mass transfer between the bubble and cloud & wake phases takes place. By allowing the volume element $\Delta V \to 0$, we get

$$\dot{n}_{bi,out} - \dot{n}_{bi,in} = \Delta \dot{n}_{bi} \tag{4.117}$$

The balance equation is transformed into an ODE

$$\frac{d\dot{n}_{bi}}{dV_b} = r_{bi}\rho_{Bb} - K_{bci}(c_{bi} - c_{ci}) \tag{4.118}$$

where the catalyst bulk density is defined by

$$\rho_{Bb} = \frac{m_{cat}}{V_b} \tag{4.119}$$

4.4.2.2 Cloud and wake phases

The cloud & wake phase remains inside the bed; thus only mass transfer is considered. The mass balance of component (i) becomes

$$K_{bci}(c_{bi} - c_{ci})\Delta V_b + r_{ci}\rho_{Bb}\Delta V_c = K_{cei}(c_{ci} - c_{ei})\Delta V_b \tag{4.120}$$

Equation (4.120) reflects, for instance, the mass transfer of a reacting component from bubble to cloud and from cloud to emulsion. After rearrangement we get,

$$K_{bci}(c_{bi} - c_{ci}) + r_{ci}\rho_{Bb}\frac{V_c}{V_b} = K_{cei}(c_{ci} - c_{ei}) \tag{4.121}$$

provided that the condition

$$\frac{\Delta V_c}{\Delta V_b} = \frac{V_c}{V_b}, \quad \text{where } \rho_{Bb} = \frac{m_{cat}}{V_c} \tag{4.122}$$

is valid. Equation (4.122) implies that the same conditions prevail everywhere in the bed.

4.4.2.3 Emulsion phase

The emulsion phase is assumed to remain inside the bed, too. Consequently, the mass balance of an arbitrary component (i) is given by

$$K_{cei}(c_{ci} - c_{ei})\Delta V_b + r_{ei}\rho_{Be}\Delta V_e = 0 \tag{4.123}$$

By letting the volume element to decrease, $\Delta V_b \to 0$, $\Delta V_e \to 0$ and $\Delta V_b/\Delta V_e = V_b/V_e$, we obtain

$$K_{cei}(c_{ci} - c_{ei}) + r_{ei}\rho_{Be}\frac{V_e}{V_b} = 0 \tag{4.124}$$

The volume ratios, such as V_e/V_b, are obtained from the corresponding void fractions (ε_b, ε_c).

Summarizing the Kunii-Levenspiel model for fluidized beds, we can conclude that $3N$ unknowns exist, namely the component concentrations for all of the three phases (c_{bi}, c_{ci}, c_{ei}). The use of the model, eqs. (4.118), (4.121) and (4.124), will be illustrated later on.

For linear kinetics, $r_i = v_i k c_i$, the Kunii-Levenspiel model can be solved analytically.

4.4.2.4 Energy balances

Energy balances are less commonly presented for fluidized beds, since the internal recirculation of small particles guarantees rather isothermal conditions inside the bed. However, in order to be able to evaluate the accumulated energy effects of the chemical reactions present, an energy balance is inevitable. A uniform temperature everywhere inside the reaction is assumed below. Thus, the phases can be treated together and we get a primary form of the energy balance:

$$\sum_j (R_{jb}\Delta V_b + R_{jc}\Delta V_c + R_{je}\Delta V_e)(-\Delta H_{rj}) = \dot{m}_b c_p \Delta T + \Delta \dot{Q} \qquad (4.125)$$

By allowing $\Delta V_b \to 0$, $\Delta T \to 0$ and $\Delta \dot{Q} \to 0$, we obtain

$$\sum_j \left(R_{jb} + R_{jc}\frac{V_c}{V_b} + R_{je}\frac{V_e}{V_b}\right)(-\Delta H_{rj})dV_b = \dot{m}_b c_p dT + d\dot{Q} \qquad (4.126)$$

Integration of eq. (4.126) formally gives

$$\int_0^{V_b} \left[\sum_j \left(R_{jb} + R_{jc}\frac{V_c}{V_b} + R_{je}\frac{V_e}{V_b}\right)\right](-\Delta H_{rj})dV_b = \dot{m}_b \int_{T_0}^{T} c_p dT + \dot{Q} \qquad (4.127)$$

The form of heat transfer term ($\dot{Q}$) depends on the actual reactor configuration. The energy balance eq. (4.127) is de facto coupled to the balances of the bubble, cloud & wake and emulsion phases. Alternatively, eq. (4.126) can be presented in the form of an ODE and solved together with the other balances for the fluidized bed.

4.4.2.5 Parameters in Kunii-Levenspiel model

Several parameters appear in the derivation of the Kunii-Levenspiel model. They are summarized here. For the catalyst bulk densities in different phases, the following relations are applied:

$$\rho_{Bb} = \frac{V_{sb}\rho_p}{V_b} = \frac{m_{catb}}{V_b} \qquad (4.128)$$

$$\rho_{Bc} = \frac{V_{sc}\rho_p}{V_b} = \frac{m_{catc}}{V_b} \qquad (4.129)$$

$$\rho_{Be} = \frac{V_{se}\rho_p}{V_b} = \frac{m_{cate}}{V_b} \qquad (4.130)$$

Traditionally, the volume of solid material in bubbles (V_{sb}), clouds (V_{sc}) and emulsion (V_{se}) are defined as:

$$\gamma_b = \frac{V_{sb}}{V_b} \qquad (4.131)$$

$$\gamma_c = \frac{V_{sc}}{V_b} \tag{4.132}$$

$$\gamma_e = \frac{V_{se}}{V_b} \tag{4.133}$$

For the mass transfer coefficients K_{bci} and K_{cei}, correlations are presented in Section 4.6.

4.5 Numerical solution of fluidized bed models

The fluidized bed model consists of the mass balance of the components in the bubble, cloud & wake and emulsion phases. If the heat effects are considerable, the energy balance equation is included, too. The mass balance for the bubble phase and the energy balance are ODEs of the type

$$\frac{dy}{dx} = f(y) \tag{4.134}$$

whereas the balances of the cloud & wake and emulsion phases are of the type

$$g(y) = 0 \tag{4.135}$$

$$h(y) = 0 \tag{4.136}$$

that is. non-linear algebraic equations (NLEs). The total number of differential equations is N in the isothermal case and $N+1$ in the non-isothermal case. The number of algebraic equations is $2N$, corresponding to the cloud & wake and emulsion phases. The total dimension of the problem is thus $3N$ or $3N+1$. Principally, we are dealing with an IVP, which can be treated by solving the bubble phase mass balance forwards with respect to the bubble phase volume (V_b) or space time (τ_b). However, the ODEs describing the bubble phase are coupled to the NLEs of the cloud & wake and emulsion phases.

Two principal numerical strategies are possible. The ODE system is considered as the main problem, and the NLEs are treated as a subproblem in the solution of the ODEs. Another alternative is that the ODEs and NLEs are treated simultaneously as a single mixed NLE-ODE problem (Figure 4.18).

Let us consider the sequential strategy first. The bubble phase concentrations are assumed to be known at the reactor inlet. The emulsion, cloud & wake phase concentrations can now be solved iteratively from the non-linear eqs. (4.121–4.124). The bubble phase balance eq. (4.118) is integrated a small step (ΔV_b) forwards, after which the NLEs of the cloud & wake and emulsion phases are solved iteratively etc.

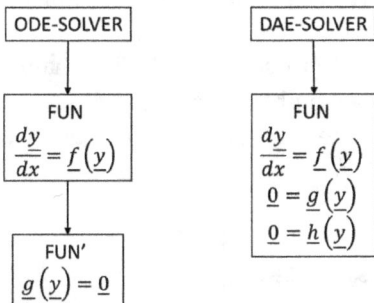

Figure 4.18: Alternative numerical strategies for solving the Kunii-Levenspiel model.

It is also possible to put a non-linear equation solver to work under the function subroutine of the ODE solver. This method is in principle more accurate than the sequential approach, but the cloud & wake and emulsion phase equations are solved a great number of times, and the function subroutine is visited each time by the ODE solver, which might unnecessarily prolong the running time. Furthermore, error control is taken over by the ODE solver focusing on the bubble phase balances only. The fact that all the balance equations are not solved with the same accuracy can be a disadvantage.

A remedy to the above-mentioned problem is to treat the whole system as a system of coupled differential-algebraic equations often abbreviated as differential-algebraic system (DAE). From the numerical solution of differential equations (Appendix A.2), the backward difference method de facto transforms the system of ODEs into a system of NLEs, since the derivatives dy/dx are described with backward differences. For instance, we take the bubble phase balance equation and use the simplest backward difference method, namely the implicit Euler method (the two-point formula), which gives

$$\frac{y_n - y_{n-1}}{\Delta x} = f(y_n) \tag{4.137}$$

where n is the index of the point. For the cloud & wake and the emulsion phase, we write

$$0 = g(y_n) \tag{4.138}$$

$$0 = h(y_n) \tag{4.139}$$

Equations (4.137–4.139) form a non-linear algebraic system originating from the ODE-NLEs, which can be solved with the aid of existing software designed for DAEs. One of the well-known software programs is DASSL designed by L. Petzold (1982). The benefit of this integrated approach is that all the equations are treated in a similar manner, under the same accuracy control. The disadvantage of the integrated approach is that starting the solution can be a problem and lead to divergence already in the beginning of the solution.

A robust, but more time-consuming approach is to solve the fully dynamic model by incorporating the accumulation term for each phase (dc/dt), to discretize the bubble phase coordinate with backward differences and to solve the large ODE system created by a robust ODE solver (Appendix A.2).

4.6 Physical properties and correlations for catalytic two-phase systems

The balance equations for catalytic two-phase reactors contain a great number of physical parameters. The most important ones among these are the diffusion coefficients, as well as the mass and heat transfer coefficients. The most common correlation equations for these quantities are discussed briefly in this section.

4.6.1 Effective diffusion coefficients in a gas phase

The effective diffusion coefficient ($D_{e,i}$) in a gas phase appears in the treatment of porous catalyst particles. The bottom line is that $D_{e,i}$ is dependent on the detailed pore structure of the catalyst particle. The simplest way to correct the diffusion coefficient is to account for the particle porosity (ε_P) and for the labyrinth structure, tortuosity of the pores (τ_P). The principle of this random pore model (RPM) is illustrated in Figure 4.19. This concept leads to the simple equation

$$D_{ei} = \frac{\varepsilon_P}{\tau_P} D_i \qquad (4.140)$$

Figure 4.19: A simple random pore model (RPM) incorporating porosity and tortuosity.

Regardless of its simplicity, eq. (4.140) is very useful in updating the diffusion coefficient (D_i) to the real conditions of a porous catalyst. For the diffusion effect itself, it is necessary to account for intermolecular collisions as well as collisions with the pore walls. The two diffusion phenomena are brought together in the expression

$$\frac{1}{D_i} = \frac{1}{D_{mi}} + \frac{1}{D_{Ki}} \tag{4.141}$$

where D_{mi} and D_{Ki} denote the molecular and Knudsen diffusion coefficients of the component (i). The former describes the intermolecular collisions, while the latter one accounts for Knudsen's diffusion; that is, molecular collisions with pore walls. Wilke and Lee (1955) have proposed an algorithm for the calculation of the molecular diffusion coefficient, starting from binary molecular diffusion coefficients,

$$D_{mi} = \frac{c - c_i}{\sum_{k=1}^{N} \frac{c_k}{D_{ik}}} \tag{4.142}$$

where $c - \sum c_i$. By introducing the mole fractions, $x_i = c_i/c$ we obtain

$$D_{mi} = \frac{1 - x_i}{\sum \frac{x_i}{D_{ik}}} \tag{4.143}$$

For the binary diffusion coefficients in a gas phase, the equation of Fuller-Schettler-Giddings (Reid et al. (1988)) has turned out to be useful

$$D_{ik} = \frac{(T/K)^{1.75} \left[\left(\frac{(g/mol)}{M_i}\right) + \left(\frac{(g/mol)}{M_k}\right)\right]^{\frac{1}{2}} \cdot 10^{-7}}{(P/atm)\left(v_i^{1/3} + v_k^{1/3}\right)^2} \; (m^2/s) \tag{4.144}$$

where the symbols are explained in Notation. For the Knudsen diffusion coefficient, we have

$$D_{Ki} = \frac{8\varepsilon_p}{3S_g \rho_p}\sqrt{\frac{2RT}{\pi M_i}} \tag{4.145}$$

where S_g is the BET surface area of the particle (expressed as () m^2/g). Liquid-phase diffusion coefficients are discussed in Chapter 6, in connection with gas-liquid reactions.

4.6.2 Mass and heat transfer coefficients around solid particles

The mass and heat transfer coefficients for gas-phase system (e.g. gas films around the catalyst particles) are expressed with semi-empirical correlations. A correlation of Wakao and Kunii (Wakao (1984)), based on an extensive collection of experimental data over various decades, is based on the relation between the dimensionless Sherwood, Schmidt and Reynolds (Re) numbers.

Wakao and Kunii have reviewed existing literature data and propose the following expression

$$Sh_p = 2 + 1.1 Sc^{1/3} Re^{0.6} \qquad (4.146)$$

where the Sherwood number for a solid particle is

$$Sh_p = \frac{k_{Gi} d_p}{D_{mi}} \quad \text{Sherwood number} \qquad (4.147)$$

and the Schmidt and Reynolds numbers are given by

$$Sc = \frac{\mu}{\rho_g D_{mi}} \quad \text{Schmidt number} \qquad (4.148)$$

$$Re = \frac{G d_p}{\mu} \quad \text{Reynolds number} \qquad (4.149)$$

In the formula above $G = \dot{m}/A$, where A is the tube cross-section.

Analogously, for heat transfer, a relation between the Nusselt number (Nu), Prandtl number (Pr) and Reynolds number (Re) is proposed:

$$Nu = 2 + 1.1 Pr^{1/3} Re^{0.6} \qquad (4.150)$$

The dimensionless numbers are defined as follows:

$$Nu = \frac{h d_p}{\lambda_g} \quad \text{Nusselt number} \qquad (4.151)$$

$$Pr = \frac{c_p \mu}{\lambda_g} \quad \text{Prandtl number} \qquad (4.152)$$

The use of eqs. (4.146–4.152) is straightforward: the Reynolds, Schmidt and Prandtl numbers are calculated, the Sherwood and Nusselt numbers are obtained, and these give the mass and heat transfer coefficients directly.

In addition, specific correlation equations have been developed for mass and heat transfer in fixed bed reactors. They take into account the porosity of the catalyst. Some correlation equations are summarized in Table 4.4.

4.6.3 Mass transfer coefficients for fluidized beds

Some correlations which are useful in the modelling of fluidized beds are summarized in Table 4.5. The bed porosity at minimum fluidization is obtained from equation (a), after which the minimum fluidization velocity can be obtained from the second-degree equation (b). The bubble rise velocity is calculated from equation (c). The diffusion coefficients for the gas phase are obtained from a suitable correlation equation, as

Table 4.4: Correlations for mass and heat transfer coefficients in fixed beds (van Santen et al. (1999)).

Mass transfer	Heat transfer	Mass transfer factor	Application domain
Gas			
$Sh = \dfrac{0.357}{\varepsilon_B} Re^{0.641} Sc^{1/3}$	$Nu = \dfrac{0.428}{\varepsilon_B} Re^{0.641} Pr^{1/3}$	$0.416 < \varepsilon < 0.788$ $j_D = 0.357 Re^{-0.359}$	$3 < Re < 2000$
Liquid			
$Sh = \dfrac{0.25}{\varepsilon_B} Re^{0.69} Sc^{1/3}$	$Nu = \dfrac{0.30}{\varepsilon_B} Re^{0.69} Pr^{1/3}$	$0.416 < \varepsilon < 0.78$ $j_D = 0.25 Re^{-0.31}$	$55 < Re < 1500$
$Sh = \dfrac{1.09}{\varepsilon_B} Re^{1/3} Sc^{1/3}$	$Nu = \dfrac{1.31}{\varepsilon_B} Re^{1/3} Pr^{1/3}$	$j_D = 1.09 Re^{-0.67}$	$0.0016 < Re < 55$

Table 4.5: Correlations for mass-transfer coefficients in fluidized beds (Levenspiel (1999)).

$\dfrac{1-\varepsilon_{mf}}{\phi'^2 \varepsilon_{mf}^2} \approx 11$	(a)
$f = \dfrac{(1-\varepsilon_{mf})^2}{\varepsilon_{mf}^3} \dfrac{a\mu}{(\phi' d_p)^2} w_{mf} + \dfrac{(1-\varepsilon_{mf})}{\varepsilon_{mf}^3} \dfrac{b\rho_G}{(\phi' d_p)^2} w_{mf}^2 = (1-\varepsilon_{mf})(\rho_p - \rho_G)g$	(b)
$K_{bci} = 4.5 \left(\dfrac{w_{mf}}{d_b} \right) + 5.85 \left(\dfrac{D_i^{1/2} g^{1/4}}{d_b^{5/4}} \right)$	(c)
$K_{cei} = 6.77 \sqrt{\dfrac{\varepsilon_{mf} D_i w_b}{d_b^3}}$	(d)

discussed in Section 4.6.1. Finally, the mass transfer coefficients needed in fluidized bed modelling can be obtained from equation (d). The readers are encouraged to familiarise themselves with literature to be updated on the most recent development of correlations in the field.

5 Modelling of three-phase systems

Catalytic three-phase reactors involve a gas phase, a liquid phase and a solid catalyst phase. Three-phase reactors are particularly important because of the existence of a certain class of processes, namely catalytic hydrogenation of organic compounds. In those cases, hydrogen always prevails in the gas phase, while the organic component to be hydrogenated resides in the liquid phase. The process is carried out in the presence or in the absence of an added solvent. A wide variety of components are hydrogenated in practise; a few examples of currently applied industrial processes are listed in Table 5.1. Basically, the hydrogenation processes imply the hydrogenation of double bonds ($C=C$), carbonyl groups ($C=O$) and aromatic rings. In addition, catalytic oxidation of carbonyl and hydroxyl groups is applied industrially. Several, often alternative reactor configurations are possible to accomplish three-phase hydrogenation.

Table 5.1: Examples of catalytic three-phase processes.

Process	Reactor type
Hydrogenation of fatty acids	slurry reactor (bubble column and stirred tank reactor)
Desulphurization	trickle bed, fluidized bed
Hydrogenated cracking	trickle bed
Fischer–Tropsch synthesis	bubble column
Hydrogenation of aromatic compounds (dearomatization), i.e. hydrogenation of benzene, toluene	trickle bed, slurry reactor
Hydrogenation of anthraquinone in the production of hydrogen peroxide	bubble column
Methanol synthesis	slurry reactor
Hydrogenation of glucose to sorbitol Hydrogenation of xylose to xylitol	slurry reactor, fixed bed reactor

Three-phase fixed bed reactors are obtained when large (on cm scale) catalyst particles are placed in beds where gas and liquid are allowed to flow through them. In the special case of a concurrent and downward laminar flow of the liquid, the fixed bed is called a trickle bed. Fixed bed reactors inevitably imply some diffusion resistance inside the catalyst pellets. A way to suppress the intraparticular resistance is to decrease the particle size (to micrometre scale) and to allow the particles to disperse in the liquid. In this case, a slurry reactor is obtained; batchwise operating slurry reactors are very popular in the production of fine and specialty chemicals. However, the catalyst has to be separated from the dispersion. In order to combine the benefits of fixed bed and slurry technologies, structured catalytic reactors have been developed. For instance in monoliths, a thin catalyst layer is precipitated on the walls of a solid (ceramic or metallic) structure, and the gas and liquid flow through the channels of the monolith. The evident advantages are the suppressed diffusion resistance and low pressure drop. An alternative way to diminish the catalyst particle size is to put the catalyst particles inside networks, which act not only as catalyst storages but also as static mixers. The different reactor configurations used for three-phase processes are illustrated in Figure 5.1.

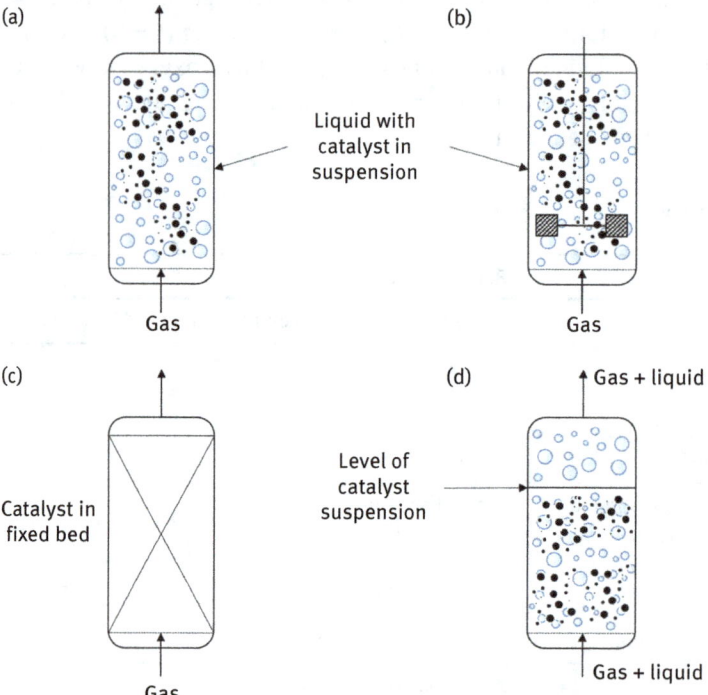

Figure 5.1: Common configurations of catalytic three-phase reactors: (a) a bubble column, (b) a tank reactor, (c) a packed bed reactor and (d) a fluidized bed reactor.

5.1 Mass balances of three-phase reactors

5.1.1 Phase boundaries

The volume element, which is essential for a catalytic three-phase reactor, is schematically illustrated in Figure 5.2; it illustrates mass and heat transfer resistances of interest: the transfer resistances on the gas–liquid interface, the liquid–solid interface as well as inside the solid catalyst. Since these are common elements for all

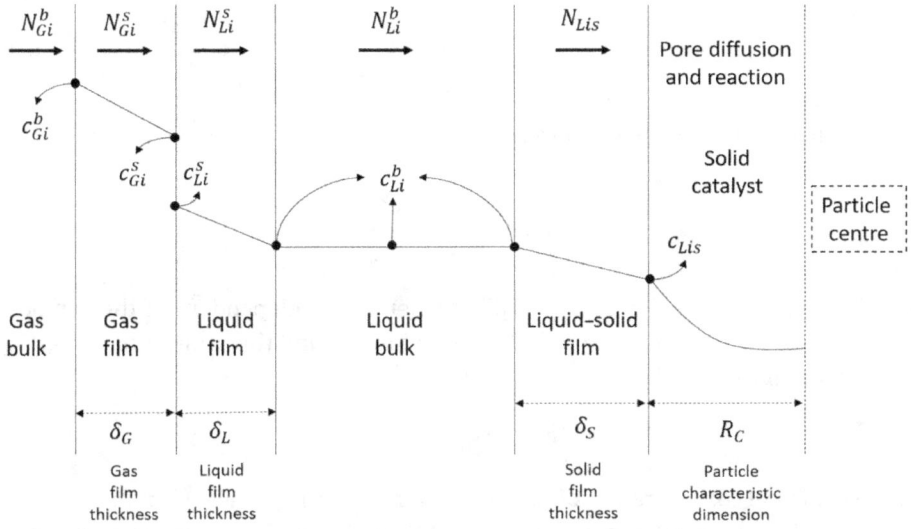

Figure 5.2: Transport effects in a volume element of a catalytic three-phase reactor.

reactor configurations, they are treated in detail in this section. For the sake of simplicity, the description of diffusion effects is based on Fick's law (Chapter 4).

The fluxes of component (i) through the gas and liquid films at the gas–liquid interface are equal, thus:

$$N_i = k_{Gi}\left(c_{Gi}^b - c_{Gi}^s\right) = k_{Li}\left(c_{Li}^s - c_{Li}^b\right) \tag{5.1}$$

where the superscripts b and s refer to bulk phase and surface concentrations (Figure 5.2).

The ratio between the gas–liquid interfacial area and the reactor volume is defined by

$$a_0 = \frac{A_{gas-liquid}}{V_R} = \frac{\Delta A}{\Delta V_R} \tag{5.2}$$

At the gas–liquid interface, a thermodynamic equilibrium is assumed to prevail. Thus, the concentrations are related by

$$K_i = \frac{c_{Gi}^s}{c_{Li}^s} \tag{5.3}$$

By utilizing the expression (5.3), $c_{Gi}^s = K_i c_{Li}^s$, the surface concentration on the liquid side c_{Li}^s can now be formally eliminated from the flux eq. (5.1):

$$c_{Li}^s = \frac{c_{Gi}^b + \frac{k_{Li}}{k_{Gi}} c_{Li}^b}{K_i + \frac{k_{Li}}{k_{Gi}}} \tag{5.4}$$

which is inserted in the flux expression given as

$$N_{Li}^b a_0 = \frac{\frac{c_{Gi}^b}{K_i} - c_{Li}^b}{\frac{1}{k_{Li} a_0} + \frac{1}{k_{Gi} a_0 K_i}} \tag{5.5}$$

Strictly speaking, eq. (5.5) is only implicit when K_i is independent of the composition. Analogously, the flux through the liquid film around the catalyst particle, N_{Lis}, can be written as

$$N_{Lis} = k_{Si}\left(c_{Li}^b - c_{Lis}\right) \tag{5.6}$$

where c_{Lis} is the concentration on the outer surface of the particle (Figure 5.2).

The ratio between the interfacial area of the particles and the reactor volume is defined by

$$a_p = \frac{A_{liquid-solid}}{V_R} = \frac{\Delta A_p}{\Delta V_R} \tag{5.7}$$

In a steady state, this flux is equal to the flux expressed with the concentration gradients on the outer surface of the catalyst particle, eq. (5.6), and, in case of no mass transfer resistance inside the particles, can even be set as equal to the generation rate of the corresponding component.

5.1.2 Liquid-phase mass balances

An axial dispersion model for the liquid bulk is considered in this section. The model is applicable to both bubble column and packed column reactors. For stirred tanks, a separate modelling effort is required. The volume elements of the gas and liquid phases will be treated separately below. The volume element is illustrated in Figure 5.3.

For an arbitrary component (i), the dynamic mass balance is written as

$$\dot{n}_{Li,in} + \left(-D_L A_L \frac{dc_{Li}}{dl}\right)_{in} + N_{Li}^b \Delta A = \dot{n}_{Li,out} + \left(-D_L A_L \frac{dc_{Li}}{dl}\right)_{out} + N_{Lis} \Delta A_p + \frac{dn_{Li}}{dt} \tag{5.8}$$

where $A_L, \Delta A, \Delta A_p$ denote the reactor cross-section, the gas–liquid interface and the liquid–solid interface, respectively.

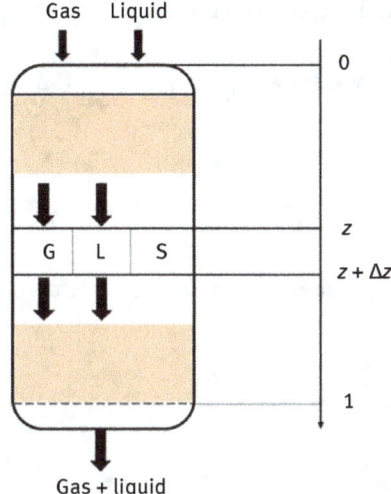

Figure 5.3: A volume element in a gas–liquid–solid column reactor.

The difference between the dispersion terms is

$$\left(D_L A_L \frac{dc_{Li}}{dl}\right)_{out} - \left(D_L A_L \frac{dc_{Li}}{dl}\right)_{in} = \Delta\left(D_L A_L \frac{dc_{Li}}{dl}\right) \quad (5.9)$$

and the difference between the plug flow contributions is

$$\dot{n}_{Li,out} - \dot{n}_{Li,in} = \Delta \dot{n}_{Li} \quad (5.10)$$

The volume element ΔV_R is defined by

$$\Delta V_R = A \Delta l \quad (5.11)$$

The surface element is defined by the liquid hold-up (ε_L) and the cross-section area (A) accordingly

$$A_L = \varepsilon_L A, \quad \varepsilon_L = \frac{V_L}{V_R} = \frac{A_L}{A} \quad (5.12)$$

The relations discussed above

$$\Delta A = a_0 \Delta V_R = a_0 A \Delta l \quad (5.13)$$

$$\Delta A_p = a_p \Delta V_R = a_p A \Delta l \quad (5.14)$$

are inserted in the balance equation, which becomes

$$\Delta\left(D_L A_L \frac{dc_{Li}}{dl}\right) + N_{Li}^b a_0 A \Delta l = \Delta \dot{n}_{Li} + N_{Lis} a_p A \Delta l + \frac{dn_{Li}}{dt} \quad (5.15)$$

The accumulation term, dn_{Li}/dt, is ($\varepsilon_L A \Delta l dc_{Li}/dt$).

Division of eq. (5.15) by the term $(A\Delta l)$ and denoting $A_L = \varepsilon_L A$ gives:

$$\frac{\Delta \dot{n}_{Li}}{\Delta l}\frac{1}{A} = \frac{\Delta}{\Delta l}\left(\varepsilon_L D_L \frac{dc_{Li}}{dl}\right) + N_{Li}^b a_0 - N_{Lis} a_p - \varepsilon_L \frac{dc_{Li}}{dt} \qquad (5.16)$$

Allowing $\Delta l \to 0$ results in

$$\frac{d\dot{n}_{Li}}{dl}\frac{1}{A} = \frac{d}{dl}\left(\varepsilon_L D_L \frac{dc_{Li}}{dl}\right) + N_{Li}^b a_0 - N_{Lis} a_p - \varepsilon_L \frac{dc_{Li}}{dt} \qquad (5.17)$$

By introducing a dimensionless coordinate, $z = l/L$, and assuming $(\varepsilon_L D_L)$ to be constant throughout the reactor, we get:

$$\frac{d\dot{n}_{Li}}{dz} = \frac{\varepsilon_L D_L V_R}{L^2}\frac{d^2 c_{Li}}{dz^2} + a_0 V_R N_{Li}^b - a_p V_R N_{Lis} - \varepsilon_L V_R \frac{dc_{Li}}{dt} \qquad (5.18)$$

The use of the approximation (see eq. (4.25))

$$\frac{d^2 c_{Li}}{dz^2} = \frac{d^2(\dot{n}_{Li}/\dot{V}_L)}{dz^2} \approx \frac{1}{\dot{V}_L}\frac{d^2 \dot{n}_{Li}}{dz^2} \qquad (5.19)$$

finally gives

$$\frac{d\dot{n}_{Li}}{dz} = \frac{\varepsilon_L D_L V_R}{\dot{V}_L L^2}\frac{d^2 \dot{n}_{Li}}{dz^2} + a_0 V_R N_{Li}^b - a_p V_R N_{Lis} - \varepsilon_L V_R \frac{dc_{Li}}{dt} \qquad (5.20)$$

The balance equation has boundary conditions. At the reactor inlet, the plug flow instantaneously changes into dispersion according to Danckwerts boundary condition:

$$\dot{n}_{Li}^0 = \dot{n}_{Li} + \left(-D_L A_L \frac{dc_{Li}}{dl}\right) \text{ at } l = 0 \qquad (5.21)$$

which is transformed to

$$\dot{n}_{Li}^0 = \dot{n}_{Li} - \frac{\varepsilon_L D_L V_R}{\dot{V}_L L^2}\frac{\dot{n}_{Li}}{dz} \text{ at } z = 0 \qquad (5.22)$$

At the outlet, the Danckwerts boundary condition (Danckwerts (1953)) requires the end of dispersion:

$$\frac{d\dot{n}_{Li}}{dz} = 0 \text{ at } z = 1 \qquad (5.23)$$

Some special cases of the initial balance equations are worth considering.
An important special case is plug flow: $D_L = 0$, and

$$\dot{n}_{Li} = \dot{n}_{Li}^0 \text{ at } z = 0 \qquad (5.24)$$

The second boundary vanishes ($z=1$), and the steady-state model of plug flow ($D_L = 0$ and $dc_{Li}/dt = 0$ in eq. (5.20)) can be solved as an initial value problem for concurrent case.

5.1.3 Gas-phase mass balances

For three-phase reactors, an analogous treatment of the gas phase is applied as for the liquid phase. However, the situation is simpler for the gas phase, because no reactions are assumed to take place in the gas phase. The transport effects are illustrated in Figure 5.2. Basically, three phenomena are involved: plug flow, axial dispersion as well as gas–liquid mass transfer. The axial dispersion is described by the term

$$-\left(D_G A_G \frac{dc_{Gi}}{dl}\right) \tag{5.25}$$

A concurrent flow of gas and liquid is assumed first, and, later on, a generalization into a countercurrent flow is presented. The mass balance thus becomes:

$$\dot{n}_{Gi,in} + \left(-D_G A_G \frac{dc_i}{dl}\right)_{in} = \dot{n}_{Gi,out} + \left(-D_G A_G \frac{dc_{Li}}{dl}\right)_{out} + N_{Gi}^b \Delta A + \frac{dn_{Gi}}{dt} \tag{5.26}$$

Through the gas and liquid films, the fluxes are equal, as indicated in the beginning of this Chapter (eq. (5.1)).

$$N_{Gi}^b = N_{Li}^b = N_{Li}^s = N_{Gi}^s \tag{5.27}$$

We introduce the following definitions,

$$\varepsilon_G = \frac{A_G}{A} \text{ and } \varepsilon_G = \frac{V_G}{V_R} \tag{5.28}$$

from which the relationships below are obtained:

$$\Delta A = a_0 \Delta V = a_0 A \Delta l = a_0 V_R \Delta z \tag{5.29}$$

The approximation for the dispersion term is introduced again,

$$\frac{d^2 c_{Gi}}{dz^2} = \frac{1}{\dot{V}_L} \frac{d^2 \dot{n}_{Gi}}{dz^2} \tag{5.30}$$

Assuming that ε_G and D_G are constant, recalling that $dn_{Gi}/dt = \varepsilon_G A \Delta z dc_{Gi}/dt$ and allowing $\Delta z \to 0$ finally gives the expression

$$\pm \frac{d\dot{n}_{Gi}}{dz} = \frac{\varepsilon_G D_G V_R}{\dot{V}_G L^2} \frac{d^2 \dot{n}_{Gi}}{dz^2} - a_0 V_R N_{Li}^b - \varepsilon_G V_R \frac{dc_{Gi}}{dt} \tag{5.31}$$

where the different signs denote:
(+) concurrent flow, and
(−) countercurrent flow. The reference flow is the liquid phase.

Boundary conditions for the gas phase can be derived in a completely analogous way as for the liquid phase (eq. 5.22). For the reactor entrance, we get:

$$\dot{n}_{Gi}^0 = \dot{n}_{Gi} \pm \frac{\varepsilon_G D_G V_R}{\dot{V}_G L^2} \frac{d\dot{n}_{Gi}}{dz} \qquad (5.32)$$

It should be remembered that eq. (5.31) is valid for
$z = 0$ for concurrent flow
$z = 1$ for countercurrent flow

At the outlet of the gas flow, the Danckwerts boundary condition gives

$$\frac{d\dot{n}_{Gi}}{dz} = 0 \qquad (5.33)$$

which again is valid for
$z = l$ for concurrent flow
$z = 0$ for countercurrent flow

In case of plug flow $D_G = 0$ and the boundary condition (5.32) are simplified to

$$\dot{n}_{Gi}^0 = \dot{n}_{Gi} \text{ at } z = 0 \text{ (concurrent) or } z = 1 \text{ (countercurrent)} \qquad (5.34)$$

The second boundary condition, eq. (5.33), vanishes for plug flow. For steady-state models, the time derivative in eq. (5.31) becomes zero.

5.1.4 Tank reactors with complete backmixing

The model for tank reactors with complete backmixing (CSTR) can be easily obtained from the general considerations presented in previous sections.
   The liquid and gas phase balances, obtained for an infinitesimal volume element, can be written as:

$$d\dot{n}_{Li} = \left(a_0 V_R N_{Li}^G - a_p V_R N_{Lis}\right)dz - \varepsilon_L V_R \frac{dc_{Li}}{dt} dz \qquad (5.35)$$

$$d\dot{n}_{Gi} = -a_0 V_R N_{Li}^G dz - \varepsilon_G V_R \frac{dc_{Gi}}{t} dz \qquad (5.36)$$

Because of uniform reactor contents, an integration of eqs. (5.35, 5.36) becomes possible

$$\int_{\dot{n}_{Li}^0}^{\dot{n}_{Li}} d\dot{n}_{Li} = \int_0^1 (a_0 V_R N_{Li}^G - a_p V_R N_{Lis}) dz - \int_0^1 \varepsilon_L V_R \frac{dc_{Li}}{dt} dz = (a_0 V_R N_{Li}^G - a_p V_R N_{Lis}) \int_0^1 dz - \varepsilon_L V_R \frac{dc_{Li}}{dt} \int_0^1 dz$$

(5.37)

$$\int_{\dot{n}_{Gi}^0}^{\dot{n}_{Gi}} d\dot{n}_{Gi} = \int_0^1 (-a_0 V_R N_{Li}^G) dz - \int_0^1 \varepsilon_G V_R \frac{dc_{Gi}}{dt} dz = (-a_0 V_R N_{Li}^G) \int_0^1 dz - \varepsilon_G V_R \frac{dc_{Gi}}{dt} \int_0^1 dz \quad (5.38)$$

and we obtain the balance equations

$$\dot{n}_{Li} - \dot{n}_{Li}^0 = (a_0 N_{Li}^G - a_p N_{Lis}) V_R - \varepsilon_L V_R \frac{dc_{Li}}{dt} \tag{5.39}$$

$$\dot{n}_{Gi} - \dot{n}_{Gi}^0 = -a_0 N_{Li}^G V_R - \varepsilon_L V_R \frac{dc_{Gi}}{dt} \tag{5.40}$$

For batch-wise operating tank reactors, the flows are zero, and the mass balances are described by the ordinary differential equations

$$\varepsilon_L \frac{dc_{Li}}{dt} = a_0 N_{Li}^G - a_p N_{Lis} \tag{5.41}$$

$$\varepsilon_G \frac{dc_{Gi}}{dt} = -a_0 N_{Li}^G \tag{5.42}$$

For all of the cases presented in this section, the initial conditions are given by

$$c_{Gi} = \frac{n_{Gi}}{V_G} = c_{Gi}^0 \text{ at } t = 0 \tag{5.43}$$

$$c_{Li} = \frac{n_{Li}}{V_L} = c_{Li}^0 \text{ at } t = 0 \tag{5.44}$$

The molar flux N_{Lis} appears in the balance equations. In the next section, we will show how it is related to the balance equations of the catalyst particles.

5.1.5 Catalyst particles in three-phase reactors

Because the catalyst particles are assumed to be completely wetted, that is, filled by a single phase (liquid), the theory presented in Chapter 4 for catalyst particles is also valid here. The flux at the catalyst outer surface is obtained from the expression

$$N_{Lis} = -\left(\rho_p R_c \int_0^1 r_i x^{a-1} dx\right) \tag{5.45}$$

as shown previously for catalyst particles. The concentration profiles needed for the calculation of the generation rates, r_i, are obtained from

$$\frac{d^2 c_i}{dx^2} + \frac{(a-1)}{x}\frac{dc_i}{dx} + \frac{\rho_p R_c^2}{D_{ei}} r_i = 0 \tag{5.46}$$

For the expression above, eq. (5.46), the following boundary conditions are applied:

$$\left(\frac{dc_i}{dx}\right)_{x=1} = \frac{k_{Si} R_c}{D_{ei}}\left(c_{Li}^b - c_{Lis}\right) \tag{5.47}$$

$$\left(\frac{dc_i}{dx}\right)_{x=0} = 0 \tag{5.48}$$

This kind of detailed treatment is necessary in the case of profound diffusion limitations inside the catalyst particle. In the absence of diffusion limitations, a considerable simplification is possible, as will be shown below.

In the absence of diffusion limitations, each generation rate, r_i, remains constant throughout the particle, and the flux expression is formally integrated.

$$N_{Lis} = -\rho_p R_c r_i \int_0^1 x^{a-1} dx = -\frac{\rho_p R_c r_i}{a} \tag{5.49}$$

Recalling that the shape factor is defined by (vide Chapter 4, Section 4.2.3),

$$a = \frac{A_p}{V_p} R_c \tag{5.50}$$

we get for all of the particles (a_p) in the volume element,

$$N_{Lis} = -\frac{\rho_p V_p}{n_p A_p} r_i = \frac{\Delta m_{cat}}{n_p A_p} r_i \tag{5.51}$$

where n_p denotes the number of particles in the volume element. The product $n_p A_p$ denotes the total interfacial area in the volume elements, that is, $n_p A_p = a_p \Delta V_R$.

The flux expression now becomes

$$N_{Lis} = -\frac{\Delta m_{cat}}{a_p \Delta V_R} r_i \tag{5.52}$$

where the catalyst bulk densities are $\rho_B = m_{cat}/V_L = \Delta m_{cat}/\Delta V_L$. The ratio $\Delta V_L/\Delta V_R$ is de facto the liquid hold-up in the volume element,

$$a_p N_{Lis} = -\varepsilon_L \rho_B r_i \tag{5.53}$$

The equation above tells us an important fact: the flux expression can be replaced by the generation rate, provided that the mass transfer resistance inside the particle is negligible. This leads to the pseudo-homogeneous model with respect to the catalyst. For instance, the axial dispersion model (eq. (5.20)) becomes:

$$\frac{d\dot{n}_{Li}}{dz} = \frac{\varepsilon_L D_L V_R}{\dot{V}_L L^2} \frac{d^2 \dot{n}_{Li}}{dz^2} + a_0 V_R N_{Li}^b + \varepsilon_L \rho_B r_i V_R \tag{5.54}$$

where the generation rate, r_i, is obtained – or more precisely, the surface concentration, c_{Lis} – by combining eqs. (5.6) and (5.53):

$$k_{Si}(c_{Li}^b - c_{Lis}) a_p = -\varepsilon_L \rho_B r_i \tag{5.55}$$

The surface concentration (c_{Lis}) is unknown and has to be eliminated from the mass balance eq. (5.55).

For linear kinetics, eq. (5.55) can be solved analytically; For instance, for a reactant (i), the generation rate is $r_i = v_i k c_{Lis}$, which is inserted in eq. (5.55), and c_{Lis} is solved with respect to c_{Li}^b. For non-linear kinetics, an iterative approach (e.g. Newton-Raphson method) is applied.

Finally, the simplest case is worth mentioning: in cases where turbulence prevails around the particles (e.g. vigorous stirring), the mass transfer resistance around the particle is eliminated. For such cases, c_{Lis} equals c_{Li}^b, and c_{Li}^b is directly used in the rate expressions (r_i).

5.1.6 Slurry reactor in the absence of mass transfer resistances

A special case of the batchwise operating three-phase reactor, a slurry reactor is of particular interest. If the catalyst particles are small enough (typically 50 μm or less), the stirring is vigorous guaranteeing the absence of external mass transfer limitations, and the gas pressure above the liquid phase is maintained constant by regulated addition of the reactant gas, the term in the liquid-phase balance equation can be replaced by eq. (5.56). At the same time, the concentration of the gas-phase component in the liquid phase is equal to the saturation concentration. Thus, the simplified mass balance becomes

$$\frac{dc_{Li}}{dt} = \rho_B r_i \tag{5.56}$$

which is formally similar to the homogeneous batch reactor model presented in Chapter 3. Equation (5.56) is frequently used in the modelling of experimentally recorded kinetic data from laboratory scale experiments. Analytical solutions obtained for homogeneous kinetics can be directly applied here, since a separation of variables and integration of eq. (5.56) gives

$$\int_{c_{0i}}^{c_i} \frac{1}{r_i} dc_{Li} = \rho_B t \tag{5.57}$$

which corresponds to the solution of homogeneous batch reactor model, accordingly:

$$\int_{c_{0i}}^{c_i} \frac{1}{r_i} dc_{Li} = t \tag{5.58}$$

Selected analytical solutions, particularly useful for catalytic systems are presented in Table 5.2. Collections of analytical solutions for homogeneous batch reactors can easily be applied here: the reaction time (t) only needs to be replaced by the product of the catalyst bulk density and the reaction time ($\rho_B t$) in the expressions. Analytical solutions for linear kinetics in various three-phase reactors are discussed in Ramachandran and Chaudhari (1983).

Table 5.2: Analytical solutions of eq. (5.58) for selected, common kinetics.

Rate equation	Mass balance	Analytical solution
$R = k$	$\frac{dc_A}{dt} = -\rho_B k$	$c_{0A} - c_A = \rho_B k t$
$R = kc_A$	$\frac{dc_A}{dt} = -\rho_B k c_A$	$\ln\left(\frac{c_{0A}}{c_A}\right) = \rho_B k t$
$R = \frac{kc_A}{1 + Kc_A}$	$\frac{dc_A}{dt} = -\rho_B \frac{kc_A}{1 + Kc_A}$	$\ln\left(\frac{c_{0A}}{c_A}\right) + K(c_{0A} - c_A) = \rho_B k t$

Observe that k can be $k = kc_{Gi}$ (the locally constant gas-phase reactant concentration) or some other, time-independent function.

5.2 Energy balances of three-phase reactors

The energy balances of three-phase reactors can be described in an analogous way to homogeneous reactors or catalytic two-phase reactors. The basic difference is that three phases appear in the system. However, the role of the gas phase is often minor in the energy balance; thus, the balance equations presented for catalytic two-phase reactors can be used as a first approximation. The addition of the energy balance equation for the gas phase in the set of equations is principally simple and does not change the structure of the mathematical problem in most cases, since three-phase reactors typically operate in a concurrent mode. If the gas and liquid phases have a vigorous contact, which is typically the case in slurry reactors, the gas and liquid phases have the same temperature, and the gas- and liquid-phase energy balances can be added together to simplify the problem. Without going into the details, the energy balance equations for a concurrent fixed bed, a CSTR and a batchwise operated slurry reactor are given below. The gas, liquid and solid phases are presumed to have the same temperature.

Fixed bed:

$$\frac{dT}{dV_R} = \frac{\sum_{j=1}^{s} R_j(-\Delta H_{rj})\rho_B \varepsilon_L - U\left(\frac{S}{V_R}\right)(T-T_c)}{\dot{m}_L c_{pL} + \dot{m}_G c_{pG}} \quad (5.59)$$

CSTR:

$$\frac{T-T_0}{V_R} = \frac{\sum_{j=1}^{s} R_j(-\Delta H_{rj})\rho_B \varepsilon_L - U\left(\frac{S}{V_R}\right)(T-T_c)}{\dot{m}_L c_{pL} + \dot{m}_G c_{pG}} \quad (5.60)$$

Batch slurry reactor:

$$\frac{dT}{dt} = \frac{\sum_{j=1}^{s} R_j(-\Delta H_{rj})\rho_B \varepsilon_L - U\left(\frac{S}{V_R}\right)(T-T_c)}{c_{pL}\rho_{0L}\varepsilon_{0L} + c_{pG}\rho_{0G}\varepsilon_{0G}} \quad (5.61)$$

5.3 Numerical aspects

The mathematical models of catalytic three-phase reactors are summarized in Table 5.3. As we can see in the Table, most cases correspond to partial and ordinary

Table 5.3: Mathematical model structures for catalytic three-phase reactors.

Tank reactor	
Dynamic	ODE(IVP)
Steady-state	NLE
Column reactor with axial dispersion	
Dynamic concurrent	PDE, parabolic system
Dynamic countercurrent	PDE, parabolic system
Steady-state concurrent	ODE(BVP)
Steady-state countercurrent	ODE(BVP)
Column reactor with plug flow	
Dynamic concurrent	PDE(hyperbolic), IVP
Dynamic countercurrent	PDE(BVP)
Steady-state concurrent	ODE(IVP)
Steady-state countercurrent	ODE(BVP)
BVP = boundary value problem	ODE = ordinary differential equation
IVP = initial value problem	PDE = partial differential equation
	NLE = non-linear equation

differential equations. The steady-state models are typically boundary value problems: the solution of these is tricky, since the inlet concentrations are known at opposite ends of the reactor for countercurrent operation. However, most three-phase

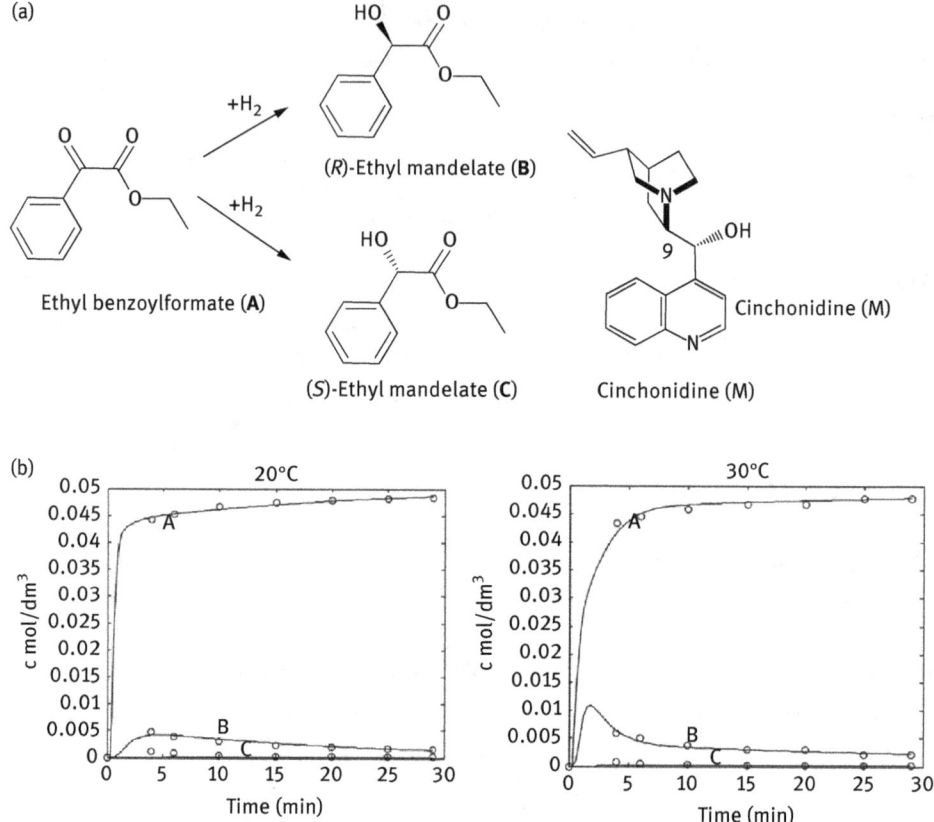

Figure 5.4: Parameter estimation in a dynamic three-phase fixed bed reactor. Case study: enantioselective hydrogenation of ethyl benzoilformate over platinum supported catalyst in the presence of a surface modifier (cinchonidine) at different temperatures. (Toukoniitty et al. 2010).

fixed bed reactors operate in a concurrent mode. The simplest model is the concurrent plug flow model in a steady-state: it is an initial value problem, which can be solved in a very straightforward manner by the algorithms for stiff ODEs (Appendix A.2). For the steady-state axial dispersion model, discretization is applied, either with finite differences or orthogonal collocation (Chapter 3). Thus, the ODEs are transformed into non-linear equations, which are solved, for instance, with the Newton-Raphson method (Appendix A.1). The Newton-Raphson method can, of course, be applied to the solution of the steady-state model for continuous tank reactors.

However, convergence problems often appear in connection with the solution of steady-state models. Therefore, the solution of dynamic models is much preferable. The spatial derivatives are discretized by finite differences or by spline collocation: as a result, the stiff PDEs are transformed into ODEs, initial value problems with respect to the reaction time. Again, we can take advantage of the very robust and efficient algorithms for stiff ODEs. Some examples are shown in Figure 5.4.

6 Modelling of gas–liquid systems

Gas–liquid reactors are used to carry out reactions between dissolved gas and components which predominantly reside in a liquid phase. From the chemical viewpoint, a gas–liquid reaction is a reaction in the liquid phase, but the mass transfer of gaseous components influences the overall rate of the process. In addition, the gaseous components have a limited solubility in the liquid phase. Theories developed for homogeneous reactors are thus not sufficient to describe gas–liquid reactors.

Industrially applied gas–liquid reactions are numerous. Gas–liquid processes have two main purposes: synthesis of chemicals through a chemical reaction and purification of gases via chemical absorption. Typical examples of manufacturing of chemicals are halogenation and sulphonation of organic components. These reactions are frequently applied as intermediate steps in the synthesis of fine chemicals. Purification of gases is a necessity in many industrial processes, for instance in the preparation of the synthesis gas for ammonia production. Moreover, removal of carbon dioxide and dihydrogen sulphide from process gases is frequently addressed. Many of the gas purification operations can be carried out as physical absorption, in the absence of chemical reactions, but the absorption rate is considerably enhanced by adding a reactive component in the liquid phase. A short list of industrially applied gas–liquid processes is provided in Table 6.1.

Table 6.1: Selected industrially applied gas–liquid reactions.

Absorption of NO_2 in H_2O in the production of HNO_3
Absorption of CO_2 in NaOH and KOH
Absorption of CO_2 in amine solutions
Absorption of CO_2 in ammonia solutions
Absorption of H_2S in amine solutions
Absorption of COS in NaOH
Oxidation of anthraquinole into anthraquinone in the H_2O_2 process
Oxidation of ethane into acetaldehyde
Oxidation of cumene into cumene hydroxide in the phenol acetone process
Oxidation of waste water
Oxidation of toluene into benzoic acid
Oxidation of xylene into phtalic acid
Chlorination of aromatic hydrocarbons
Chlorination of acetic acid into monochloroacetic acid

A large variety of gas–liquid reactors is used on the industrial scale. Some typical reactor configurations are displayed in Figure 6.1. For chemical synthesis on a small scale, a semi-batch stirred tank is frequently used. Fresh feed gas is bubbled through the liquid phase, and unreacted gas is circulated back into the reactor vessel. Stirred tanks can be operated continuously (CSTRs (Continuous Stirred Tank Reactors)), but this is not particularly efficient, because the conversion level remains low. For very special reaction kinetics, such as autocatalytic processes, a CSTR, can become a preferable choice. The efficiency of a CSTR can be improved by coupling tanks in series, similarly to homogeneous processes. The gas-phase component can be fed separately to the tanks in order to steer the product selectivity.

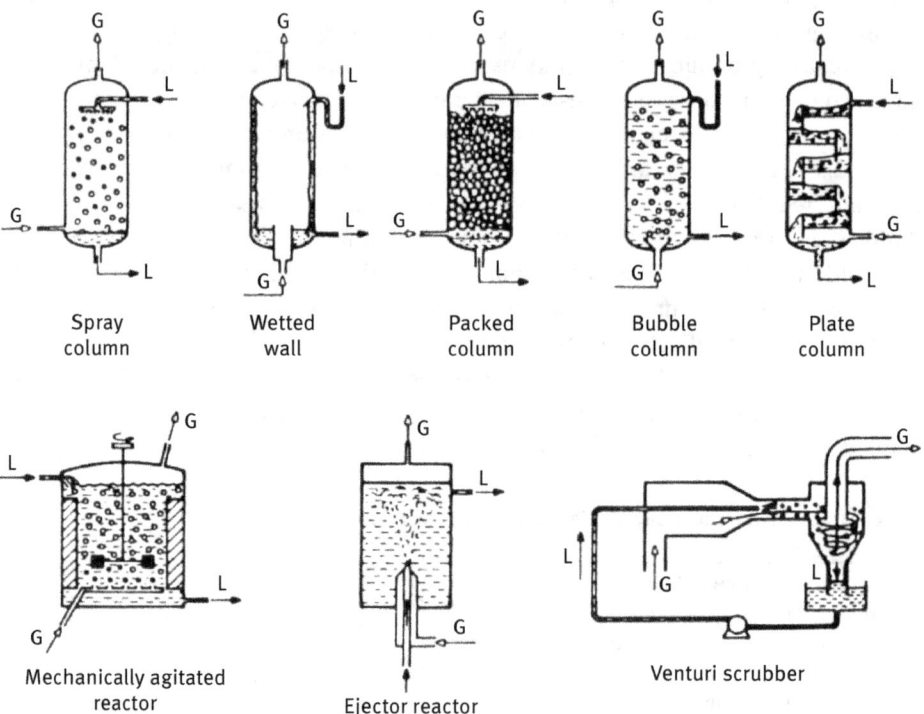

Figure 6.1: Typical, industrially applied reactor configurations and column packing materials (Charpentier (1981)).

The most common gas–liquid reactor configuration is a bubble column (Figure 6.1). Its operation principle is very simple: gas and liquid flow through a column, which might be equipped with static mixers. Bubble columns can also operate as semi-batch reactors; typically the liquid phase is in batch, and a gas flows through the reactor. In this way, slow organic reactions are carried out in an efficient way.

The reactors discussed above are used in the synthesis of chemicals. In the purification of gases, the requirements for an efficient reactor system are rather different. The reactions are often of ionic character, such as absorption of CO_2 in KOH and absorption of H_2S into amine solutions. This implies that the intrinsic process is rapid, but very low gas contents are required at the reactor outlet. Thus, an efficient gas–liquid contact, that is, a high interfacial area-to-reactor volume ratio is demanded. This is accomplished by filling the column with a packing material (Figure 6.2). The packed column reactor is operated in a countercurrent mode. This ensures that the outlet gas is in contact with a fresh liquid which has a high concentration of the liquid-phase component.

Figure 6.2: Common types of column packing materials (Trambouze et al.(1988)).

In spite of the large variety of physical configurations of gas–liquid reactors, the modelling principles can be very much generalized, as will be demonstrated in the subsequent sections. The central issue is the description of the gas–liquid contact in the interfacial domain of the reactor. During the history of chemical engineering, several theories have been developed to describe the interfacial phenomena of gas–liquid reactions.

6.1 Gas–liquid contact

It is generally agreed that the domains close to the interface of a gas bubble and the liquid phase deviate from the bulk phases of gas and liquid, particularly in reactors operating on an industrial scale. In laboratory reactors, an enormous stirring effect per reactor volume can be achieved, and turbulence is locally well-developed. Consequently, the concentration of dissolved gas approaches the solubility limit of the gas. For reactors operating on a large scale, this is not often possible, since the stirring effect is insufficient. Instead, stagnant zones develop, particularly on the liquid-side of the interface (Figure 6.3).

The simplest way to describe this effect is provided by the two-film theory developed in early 1900s. According to the film theory, the gas–liquid interface is surrounded by gas and liquid films, in which the mass transport takes place solely via

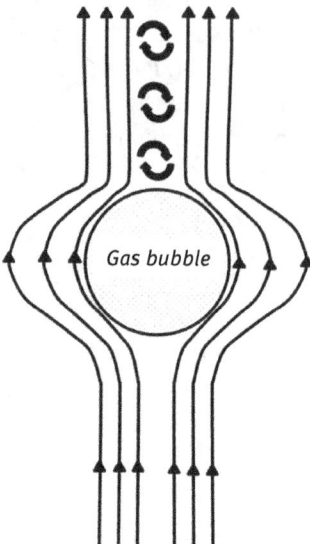

Figure 6.3: Flow profiles surrounding a gas bubble.

molecular diffusion. Thus, the theory developed for simultaneous reaction and diffusion in porous media can be applied directly to the liquid films in gas–liquid contact (Section 4.1). For the gas film, the situation is simpler, because no reactions take place in the gas phase. The main weakness of the film theory is the description of the film thickness, which is not an unequivocal quantity: the stagnant zone changes gradually to a turbulent zone in the bulk phase.

Because of the theoretical pitfalls of the film theory, more advanced concepts have been developed for the gas–liquid contact, such as the penetration theory of Higbie (1935) and the surface renewal theory of Danckwerts (Danckwerts (1970)). Both of these consider dynamic effects in the liquid phase, namely transient diffusion effects. The vicinity of the gas–liquid interface is considered as an infinite continuum where reaction and diffusion take place simultaneously. The theory of Danckwerts describes the surface as a mosaic structure where the surface elements follow a residence time distribution, typically an exponential function of the type e^{-st} (s = surface renewal parameter). The time-dependent flux ($N(t)$) is obtained from the solution of the dynamic mass balance of the liquid phase, for different residence times on the interface. The average flux is finally calculated by integration of the time-dependent flux.

Numerical comparisons have shown that fluxes predicted by the surface renewal theory and the simple film theory are rather close. This might be the reason for the fact that the film theory still competes very well with the more advanced theories. As a matter of fact, the breakthrough of numerical computing has even strengthened the role of the two-film theory, despite its philosophical weaknesses. Below, we will limit the quantitative treatment to the film theory.

6.2 Gas and liquid films

At the core of understanding gas–liquid reactors is the description of the gas–liquid contact in the interfacial area; therefore, the balance equations of gas and liquid films are treated in this section. Chemical reactions are assumed to proceed in the liquid film, while only mass transfer appears in the gas film. Thus, the description of the gas film can be obtained as a special case of the liquid film solution.

6.2.1 Mass balances for films

The gas–liquid film in any gas–liquid reactor can be described by the schematic representation displayed in Figure 6.4.

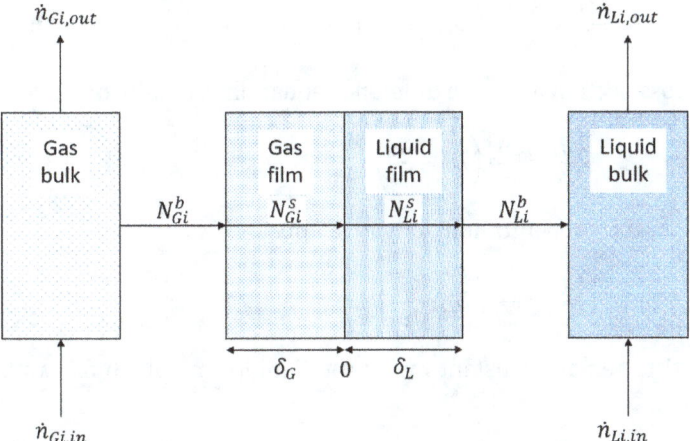

Figure 6.4: A schematic image of a gas–liquid film in a gas–liquid reactor.

A volume element in the liquid film is defined as

$$\Delta V_L = A \Delta z \tag{6.1}$$

where A and Δz denote the cross-section and length, respectively. The mass balance of an arbitrary component (i) is based on the following principle:

Influx through diffusion + reaction = outflux through diffusion + accumulation

Quantitatively this can be expressed by

$$\left(-D_{Li}\frac{dc_{Li}}{dz}A\right)_{in} + r_i \Delta V_L = \left(-D_{Li}\frac{dc_{Li}}{dz}A\right)_{out} + \frac{dn_{Li}}{dt} \qquad (6.2)$$

The accumulation term is expressed by the concentrations in the volume element:

$$\frac{dn_{Li}}{dt} = \Delta V_L \frac{dc_{Li}}{dt} = A\Delta z \frac{dc_{Li}}{dt} \qquad (6.3)$$

The difference between the diffusion terms is denoted by

$$\left(D_{Li}\frac{dc_{Li}}{dz}A\right)_{out} - \left(D_{Li}\frac{dc_{Li}}{dz}A\right)_{in} = \Delta\left(D_{Li}\frac{dc_{Li}}{dz}A\right) \qquad (6.4)$$

Thus, the balance for the liquid film becomes

$$\frac{dc_{Li}}{dt} = \frac{\Delta}{A\Delta z}\left(D_{Li}\frac{dc_{Li}}{dz}A\right) + r_i \qquad (6.5)$$

After eliminating the cross-section area, the difference equation is obtained:

$$\frac{dc_{Li}}{dt} = \frac{\Delta}{\Delta z}\left(D_{Li}\frac{dc_{Li}}{dz}\right) + r_i \qquad (6.6)$$

Letting Δz approach zero, the concentration derivative appears:

$$\frac{dc_{Li}}{dt} = \frac{d}{dz}\left(D_{Li}\frac{dc_{Li}}{dz}\right) + r_i \qquad (6.7)$$

Usually, D_{Li} is assumed to remain constant inside the liquid film, which allows a simplification of eq. (6.7):

$$\frac{dc_{Li}}{dt} = D_{Li}\frac{d^2 c_{Li}}{dz^2} + r_i \qquad (6.8)$$

Introduction of a dimensionless coordinate (x) for the liquid film,

$$z = x\delta_L \qquad (6.9)$$

where δ_L denotes the film thickness, implies that

$$\frac{dc_{Li}}{dt} = \frac{D_{Li}}{\delta_L^2}\frac{d^2 c_{Li}}{dx^2} + r_i \qquad (6.10)$$

According to the film theory, the mass transfer coefficient (k_{Li}) is defined by

$$k_{Li} = \frac{D_{Li}}{\delta_L} \qquad (6.11)$$

In a steady state, the time derivative disappears,

$$\frac{dc_{Li}}{dt} = 0 \tag{6.12}$$

and we obtain the film equation in the following forms

$$D_{Li}\frac{d^2 c_{Li}}{dz^2} + r_i = 0, \qquad z \in [0, \delta_L] \tag{6.13}$$

$$\frac{D_{Li}}{\delta_L^2}\frac{d^2 c_{Li}}{dx^2} + r_i = 0, \qquad x \in [0, 1] \tag{6.14}$$

The latter equation represents the dimensionless length coordinate.

A completely analogous derivation can be carried out for the gas film. However, since no chemical reactions take place, a simpler equation is obtained, analogously with eq. (6.10),

$$\frac{dc_{Gi}}{dt} = \frac{D_{Gi}}{\delta_G^2}\frac{d^2 c_{Gi}}{dx^2}, \qquad x \in [0, 1] \tag{6.15}$$

The mass transfer coefficient for the gas film is defined by

$$k_{Gi} = \frac{D_{Gi}}{\delta_G} \tag{6.16}$$

The film equations derived above have 2N boundary conditions (BC). The BCs at the gas–liquid interface are obtained by setting the fluxes equal on the interface:

$$N_{Gi}^s = N_{Li}^s \tag{6.17}$$

where the fluxes according to Fick's law are expressed as

$$N_{Gi}^s = -D_{Gi}\left(\frac{dc_{Gi}}{dz}\right) = -\frac{D_{Gi}}{\delta_G}\left(\frac{dc_{Gi}}{dx}\right) \tag{6.18}$$

$$N_{Li}^s = -D_{Li}\left(\frac{dc_{Li}}{dz}\right) = -\frac{D_{Li}}{\delta_L}\left(\frac{dc_{Li}}{dx}\right) \tag{6.19}$$

In a steady state, the time derivative of the concentration is zero, and the flux through the gas film is obtained from

$$\frac{D_{Gi}}{\delta_G^2}\frac{d^2 c_{Gi}}{dx^2} = 0 \tag{6.20}$$

which gives linear concentration profiles. The solution is

$$N_{Gi}^b = N_{Gi}^s = \frac{D_{Gi}}{\delta_G}\left(c_{Gi}^b - c_{Gi}^s\right) \tag{6.21}$$

Consequently, we have at $x = 0$:

$$\frac{D_{Gi}}{\delta_G}\left(c_{Gi}^b - c_{Gi}^s\right) = -\frac{D_{Li}}{\delta_L}\left(\frac{dc_{Li}}{dx}\right)_{x=0} \tag{6.22}$$

The use of the phase-equilibrium ratio (K_i) on the gas–liquid interface implies that the gas-phase concentration can be easily eliminated, as described below:

$$K_i = \frac{c_{Gi}^s}{c_{Li}^s}, \quad c_{Gi}^s = K_i c_{Li}^s \tag{6.23}$$

Equation (6.23) is inserted in eq. (6.22), giving

$$\frac{D_{Li}}{\delta_L}\left(\frac{dc_{Li}}{dx}\right)_{x=0} + \frac{D_{Gi}}{\delta_G}\left(-K_i c_{Li}^s + c_{Gi}^b\right) = 0 \tag{6.24}$$

A rearrangement of eq. (6.24) gives

$$\frac{D_{Li}\delta_G}{D_{Gi}\delta_L}\left(\frac{dc_{Li}}{dx}\right)_{x=0} + \left(c_{Gi}^b - K_i c_{Li}^s\right) = 0 \tag{6.25}$$

$$\frac{D_{Li}}{\delta_L}\frac{\delta_G}{D_{Gi}} = \frac{k_{Li}}{k_{Gi}}, \quad c_{Li}^s = c_{Li} \tag{6.26}$$

where the ratio between the mass transfer coefficients appears spontaneously:

$$\frac{k_{Li}}{k_{Gi}}\left(\frac{dc_{Li}}{dx}\right)_{x=0} + c_{Gi}^b - K_i c_{Li} = 0, \quad \text{at } x = 0 \tag{6.27}$$

The BC is thus compressed to eq. (6.27), which is a very general BC, valid at $x = 0$. Some special cases will be discussed below.

In the absence of the gas-film resistance, $k_{Gi} \to \infty$, the first term in eq. (6.27) vanishes. Towards the end of the film ($x = 1$, $z = \delta_L$), the BC is

$$c_{Li} = c_{Li}^b, \quad x = 1 \tag{6.28}$$

that is, the concentration is locally known.

For non-volatile compounds, we have a special case of eq. (6.27), since c_{Gi}^b and K_i are zero.

$$\frac{dc_{Li}}{dx} = 0, \quad x = 0 \tag{6.29}$$

An analytical solution of the film equations is possible for simple kinetics, for example, isothermal first- and zero-order kinetics. Some analytical solutions are provided in Table 6.2. For general kinetics, a numerical solution is inevitable.

Table 6.2: Analytical solutions of film equations for zero- and first-order isothermal cases $\left(M^{1/2} = \text{Hatta number}\right)$.

Zero-order reaction

$$c_{LA}(z) = -\frac{v_A k \delta_L^2}{2 D_{LA}} \left(\frac{z}{\delta_L}\right)^2 + \left(c_{LA}^b - c_{LA}^s + \frac{v_A k \delta_L^2}{2 D_{LA}}\right)\left(\frac{z}{\delta_L}\right) + c_{LA}^s$$

First-order reaction

$$c_{LA}(z) = \frac{c_{LA}^b \sinh\left(M^{1/2}\left(\frac{z}{\delta_L}\right)\right) + c_{LA}^s \sinh\left(M^{1/2}\left(1-\frac{z}{\delta_L}\right)\right)}{\sinh\left(M^{1/2}\right)}$$

$$M^{1/2} = \delta_L \sqrt{-\frac{v_A k}{D_{LA}}}$$

6.2.2 Energy balances for liquid films

In cases where the reaction enthalpy deviates clearly from zero, energy balances for liquid films are needed to predict the temperature profile inside the liquid film as well as to describe the heat flux between gas and liquid phases. Several chemical reactions are assumed to take place in the liquid film. Furthermore, heat transfer inside the film is described by the law of Fourier. The energy balance for the liquid film can, therefore, be written as

$$\left(-\lambda_L \frac{dT}{dz} A\right)_{in} + \sum_j R_j\left(-\Delta H_{rj}\right)\Delta V_L = \left(-\lambda_L \frac{dT}{dz} A\right)_{out} + \sum_i c_{Li} c_{vmLi} \Delta V_L \frac{dT}{dt} \quad (6.30)$$

After recalling that $\Delta V_L = A\Delta z$, we get

$$\frac{\Delta}{\Delta z}\left(\lambda_L \frac{dT}{dz}\right) + \sum_j R_j\left(-\Delta H_{rj}\right) = \sum_i c_{Li} c_{vmLi} \frac{dT}{dt} \quad (6.31)$$

Allowing $\Delta z \to 0$ and assuming λ_L . constant, which is a very reasonable approximation for liquids, the second-order differential equation is obtained:

$$\lambda_L \frac{d^2 T}{dz^2} + \sum_j R_j\left(-\Delta H_{rj}\right) = \sum_i c_{Li} c_{vmLi} \frac{dT}{dt} \quad (6.32)$$

Introduction of a dimensionless coordinate gives ($x = z/\delta_L$)

$$\sum_i c_{Li} c_{vmLi} \frac{dT}{dt} = \frac{\lambda_L}{\delta_L^2} \frac{d^2T}{dx^2} + \sum_j R_j(-\Delta H_{rj}) \tag{6.33}$$

Typically, for liquid-phase systems, the difference between c_{vmLi} and c_{pmLi} is minor, and the term $\sum c_{Li} c_{vmLi}$ can be replaced by the mass-based heat capacity and density, $c_p \rho$.

In a steady state, the time derivative vanishes and the energy balance is reduced to

$$\frac{\lambda_L}{\delta_L^2} \frac{d^2T}{dx^2} + \sum_j R_j(-\Delta H_{rj}) = 0 \tag{6.34}$$

A heat transfer coefficient h_L is defined analogously with the mass transfer coefficient,

$$h_L = \frac{\lambda_L}{\delta_L} \tag{6.35}$$

The BC, on the gas–liquid interface, is obtained by setting the heat fluxes equal on the interface:

$$M_G^s = h_G(T_G^b - T_G^s), \quad h_G = \frac{\lambda_G}{\delta_G} \tag{6.36}$$

$$M_L^s = -\frac{\lambda_L}{\delta_L}\left(\frac{dT}{dx}\right)_{x=0} = -h_L\left(\frac{dT}{dx}\right)_{x=0} \tag{6.37}$$

We obtain

$$-h_L\left(\frac{dT}{dx}\right)_{x=0} + h_G(T_G^b - T_G^s) = 0, \quad T_G^s = T \tag{6.38}$$

$$\frac{h_L}{h_G}\left(\frac{dT}{dx}\right)_{x=0} + T - T_G^b = 0, \quad x = 0 \tag{6.39}$$

which is completely analogous with the mass balance equation, eq. (6.27). At the end of the film, we have the simple BC,

$$T = T_L^b, \quad x = 1 \tag{6.40}$$

since the temperature is locally known.

The mathematical model for non-isothermal cases thus consists of mass balances, eqs. (6.10) and (6.15), along with the energy balance equation, eq. (6.33). Since the rate equations depend exponentially on the temperature, the system of differential equations is strongly non-linear also for zero- and first-order kinetics. Concentration and temperature profiles in the liquid film are obtained through numerical solution. The liquid film, however, represents a local quantity only. In order to get a global view of reactor performance, the film equations are coupled to macroscopic balance equations of a reactor. Therefore, gas–liquid tank reactors and column reactors are considered in the following sections.

6.3 Gas–liquid tank reactors

The mass balances for completely backmixed, gas–liquid tank reactors are described in this section. As a special case, semi-batch reactors are considered. In addition, the coupling of the film and bulk phase balance equations is discussed. A schematic view of the reactor is provided in Figure 6.5.

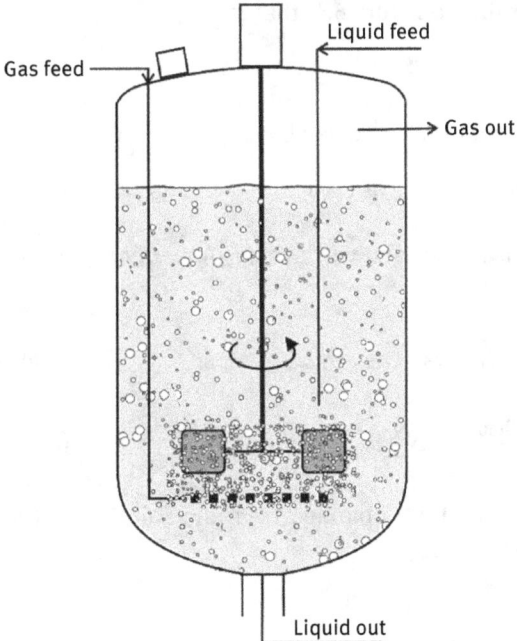

Figure 6.5: A schematic illustration of a gas–liquid tank reactor.

For a continuous tank reactor, the mass balance of an arbitrary component (i) is given as the general conservation equation for the liquid phase:

$$\dot{n}_{OLi} + N^G_{Li}A + V_L r_i = \dot{n}_{Li} + \frac{dn_{Li}}{dt} \tag{6.41}$$

$$V_L = \varepsilon_L V_R \tag{6.42}$$

where ε_L denotes the liquid hold-up; for reactors without packing elements, $\varepsilon_G + \varepsilon_L = 1$. For the gas phase, the balance equation is simpler, since chemical reactions are absent:

$$\dot{n}_{0Gi} = N_{Gi}^b A + \dot{n}_{Gi} + \frac{dn_{Gi}}{dt} \tag{6.43}$$

For the interfacial area, we introduce $A = a_0 V_R$ and obtain, after some rearrangement:

$$\frac{dn_{Li}}{dt} = \dot{n}_{0Li} - \dot{n}_{Li} + N_{Li}^b a_0 V_R + (1 - \varepsilon_G) V_R r_i \tag{6.44}$$

$$\frac{dn_{Gi}}{dt} = \dot{n}_{0Gi} - \dot{n}_{Gi} + N_{Gi}^b a_0 V_R \tag{6.45}$$

The interfacial fluxes are related according to Figure 6.4, that is

$$N_{Gi}^b = N_{Li}^s \neq N_{Li}^b \tag{6.46}$$

The initial conditions for the balance equations are given by

$$n_{Li} = n_{Li}^0 \text{ and } n_{Gi} = n_{Gi}^0, \text{ at } t = 0 \tag{6.47}$$

In a steady state, the time derivatives disappear, and we have $dn_{Li}/dt = 0$, $dn_{Gi}/dt = 0$. The balance equations take very simple algebraic forms:

$$\frac{\dot{n}_{Li} - \dot{n}_{Li0}}{V_R} = N_{Li}^b a_0 + (1 - \varepsilon_G) r_i \tag{6.48}$$

$$\frac{\dot{n}_{Gi} - \dot{n}_{Gi0}}{V_R} = -N_{Gi}^b a_0 \tag{6.49}$$

The fluxes needed in eqs. (6.48) and (6.49) are obtained from the solution of the film equation, as discussed in the previous section.

6.4 Gas–liquid column reactors

In the derivation of mass balances for gas–liquid column reactors, the following basic assumptions are made: axial dispersion effects are included both for gas and liquid and the reactor can operate in a concurrent or a countercurrent mode. Radial dispersion is neglected. The considerations are mainly devoted to columns without packing materials, but the treatment can easily be extended to packed columns, too.

In column reactors, the degree of backmixing is very different in the gas and liquid phases. A general rule of thumb is that the gas phase is close to plug flow, while the degree of backmixing can be high for the liquid phase (Deckwer (1985)).

An infinitesimal volume element of the column is considered (Figure 6.6). The coordinate system is selected according to the flow of liquid.

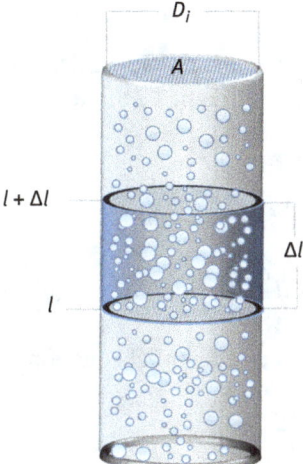

Figure 6.6: A gas-liquid column reactor.

For the liquid phase, the balance equation is written as:

$$\dot{n}_{Li,in} + \left(-D_L \frac{dc_{Li}}{dl}\right)_{in} A_L + N_{Li}^b \Delta A + r_i \Delta V_L = \dot{n}_{Li,out} + \left(-D_L \frac{dc_{Li}}{dl}\right)_{out} A_L + \frac{dn_{Li}}{dt} \quad (6.50)$$

where the interfacial area is defined by

$$A_L = \varepsilon_L A, \qquad \Delta V_L = \varepsilon_L A \Delta l \quad (6.51)$$

The definition is inserted in eq. (6.50), which becomes

$$\Delta\left(D_L \frac{dc_{Li}}{dl}\right)\varepsilon_L A + N_{Li}^b \Delta A + r_i \varepsilon_L A \Delta l = \Delta \dot{n}_{Li} \quad (6.52)$$

The interfacial element, ΔA, is obtained from

$$a_v = \frac{\Delta A}{\Delta V_R} \quad (6.53)$$

In literature, correlations are provided for a_0. At the next stage, the volume and length elements are allowed to diminish, $\Delta l \to 0$, and the differential equation is obtained:

$$\frac{d}{dl}\left[\left(D_L \frac{dc_{Li}}{dl}\right)\varepsilon_L\right] + N_{Li}^b a_v + r_i \varepsilon_L = \frac{d\dot{n}_{Li}}{Adl} \tag{6.54}$$

After introducing a dimensionless variable, $z = l/L$, the liquid-phase balance equation gets the final form:

$$\frac{d\dot{n}_{Li}}{dz} = \frac{D_L \varepsilon_L V_R}{L^2} \frac{d^2 c_{Li}}{dz^2} + N_{Li}^b a_v V_R + r_i \varepsilon_L V_R \tag{6.55}$$

For the gas phase – in the volume element – the balance equations can be derived in completely analogous manner, as illustrated below. The difference equations for the gas-phase volume element become:

$$\dot{n}_{Gi,in} + \left(-D_G \frac{dc_{Gi}}{dl}\right)_{in} A_G = \dot{n}_{Gi,out} + \left(-D_G \frac{dc_{Gi}}{dl}\right)_{out} A_G + N_{Gi}^b \Delta A + \frac{dn_{Gi}}{dt} \tag{6.56}$$

$$\Delta\left[\left(D_G \frac{dc_{Gi}}{dl}\right)\varepsilon_G A\right] = \Delta \dot{n}_{Gi} + N_{Gi}^b A \Delta l + A \Delta l \frac{dc_{Gi}}{dt} \tag{6.57}$$

After letting the volume element to diminish, $\Delta l \to 0$, we get a differential equation:

$$\frac{d}{dl}\left(D_G \frac{dc_{Gi}}{dl}\right)\varepsilon_G A - N_{Gi}^b a_v A = \frac{d\dot{n}_{Gi}}{dl} + \frac{dc_{Gi}}{dt} A \tag{6.58}$$

Assuming steady-state operation, the dimensionless coordinate $z = l/L$ is introduced,

$$\frac{d\dot{n}_{Gi}}{dz} = \frac{D_G \varepsilon_G V_R}{L^2} \frac{d^2 c_{Gi}}{dz^2} - N_{Gi}^b a_v V_R \tag{6.59}$$

The second derivative of the concentration can be approximated by

$$\frac{d^2 c_{Gi}}{dz^2} = \frac{d^2(\dot{n}_{Gi}/\dot{V}_G)}{dz^2} \approx \frac{1}{\dot{V}_G} \frac{d^2 \dot{n}_{Gi}}{dz^2} \tag{6.60}$$

which, in turn, is inserted in eq. (6.56). As a result, we obtain everything as a function of the molar flow:

$$\frac{d\dot{n}_{Gi}}{dz} = \frac{D_G \varepsilon_G V_R}{L^2 \dot{V}_G} \frac{d^2 \dot{n}_{Gi}}{dz^2} - N_{Gi}^b a_v V_R \tag{6.61}$$

The derivations above were presented in a coordinate system defined by the direction of the gas flow. For a concurrent flow, the gas and liquid coordinates coincide, while for a countercurrent flow z is changed to $1 - z$. In the steady-state balance equation,

$$\pm \frac{d\dot{n}_{Gi}}{dz} = \frac{D_G \varepsilon_G V_R}{L^2 \dot{V}_G} \frac{d^2 \dot{n}_{Gi}}{dz^2} - N_{Gi}^b a_v V_R \tag{6.62}$$

the signs + and − refer to concurrent and countercurrent flows, respectively. The Péclet number for the gas phase is defined by

$$\frac{L^2 \dot{V}_G}{D_G V_R} = \frac{L^2 \dot{V}_G}{D_G A L} = \frac{w_G L}{D_G} = Pe_G = Bo_G \qquad (6.63)$$

For plug-flow conditions, the balance equations are essentially simplified, since the dispersion terms are ignored:

$$\pm \frac{d\dot{n}_{Gi}}{dz} = -N^b_{Gi} a_v V_R \qquad (6.64)$$

$$\frac{d\dot{n}_{Gi}}{dz} = -N^b_{Gi} a_v V_R \quad \text{cocurrent} \qquad (6.65)$$

$$\frac{d\dot{n}_{Li}}{dz} = \frac{D_L \varepsilon_L V_R}{L^2} \frac{d^2 c_{Li}}{dz^2} + N^b_{Li} a_v V_R + \varepsilon_L V_R r_i \qquad (6.66)$$

$$\frac{d\dot{n}_{Gi}}{dz} = +N^b_{Gi} a_v V_R \quad \text{countercurrent} \qquad (6.67)$$

The dispersion term is approximated by

$$\frac{d^2 c_{Li}}{dz^2} = \frac{d^2 (\dot{n}_{Li}/\dot{V}_L)}{dz^2} \approx \frac{1}{\dot{V}_L} \frac{d^2 \dot{n}_{Li}}{dz^2} \qquad (6.68)$$

after which the balance equation takes the form ($\varepsilon_L = 1 - \varepsilon_G$):

$$\frac{d\dot{n}_{Li}}{dz} = \frac{(1-\varepsilon_G) D_L V_R}{L^2 \dot{V}_L} \frac{d^2 \dot{n}_{Li}}{dz^2} + N^b_{Li} a_v V_R + (1-\varepsilon_G) V_R r_i \qquad (6.69)$$

where a dimensionless quantity, the liquid-phase Péclet number appears spontaneously,

$$\frac{L^2 \dot{V}_L}{D_L V_R} = \frac{w_L L}{D_L} = Pe_L = Bo_L \qquad (6.70)$$

The dimensionless number is usually called the Péclet number (*Pe*), although, in German-speaking countries, it is often referred to as the Bodenstein number (*Bo*). The following limiting cases are of interest: $Pe_L = 0$ (high value of the dispersion coefficient) implies complete backmixing, while $Pe_L = \infty$ gives the plug flow model:

$$\frac{d\dot{n}_{Li}}{dz} = N^b_{Li} a_v V_R + (1-\varepsilon_G) V_R r_i \qquad (6.71)$$

6.4.1 Boundary conditions for balance equations

The second-order differential equations describing gas and liquid phases form a boundary value problem (BVP), in which $2N$ (N = number of components) boundary conditions (BCs) are necessary. The BC for the entrance of the liquid phase is obtained in the following consideration: a molar flow (plug flow) is fed into the reactor vessel and instantaneously transformed into a dispersed flow. This leads to the relationship,

$$\dot{n}_{Li}^0 = \dot{n}_{Li} + dispersion \tag{6.72}$$

$$\dot{n}_{Li}^0 = \dot{n}_{Li} + \left(-D_L \frac{dc_{Li}}{dl} A_L\right)_{l=0} \tag{6.73}$$

which can be rewritten as ($A_L = \varepsilon_L A$, $A = V_R/L$)

$$\dot{n}_{Li}^0 = \dot{n}_{Li} - \frac{\varepsilon_L D_L V_R}{L^2 \dot{V}_L} \frac{d\dot{n}_{Li}}{dz}, \qquad z = 0 \tag{6.74}$$

The above eq. (6.74) is the well-known Danckwerts' BC for the reactor inlet.

At the outlet, all concentration and molar flow gradients are presumed to disappear:

$$\frac{d\dot{n}_{Li}}{dz} = 0, \qquad z = L \tag{6.75}$$

This is the thermodynamically consistent outlet BC proposed by Danckwerts (1953).

For the gas phase, a completely analogous reasoning gives us

$$\dot{n}_{Gi}^0 = \dot{n}_{Gi} \pm \frac{\varepsilon_G D_G V_R}{L^2 \dot{V}_G} \frac{d\dot{n}_{Gi}}{dz} \tag{6.76}$$

where the signs, $-$ and $+$, again denote cocurrent and countercurrent flows, respectively. For the outlet we have, in line with eq. (6.75)

$$\frac{d\dot{n}_{Gi}}{dz} = 0 \tag{6.77}$$

where $z = 1$ for cocurrent and $z = 0$ for countercurrent operation.

6.5 Energy balances for gas–liquid reactors

The energy balances for gas–liquid reactors can be derived in an analogous manner to mass balances. The role of the gas phase in the energy balance is usually minor, and it is thus ignored in many treatments. If the gas and liquid phases can be assumed to have the same temperature, the treatment is essentially simplified. However, it is necessary to take into account the heat effect originating from the film reactions if the reactions are rapid. Thus, two heat source terms are in principle included, those originating from film and bulk reactions. In many cases, one of them is negligible. For slow reactions, only the bulk-phase source term is of importance. The steady-state energy balances for column reactors and perfectly back-mixed tank reactors (CSTRs) are given below.

Column reactor:

$$\frac{dT}{dV_R} = \frac{\int_0^{\delta_L} \sum_{j=1}^{s} R_j(-\Delta H_{rj})dza_v + \sum_{j=1}^{s} R_j(-\Delta H_{rj})\varepsilon_L - U\left(\frac{S}{V_R}\right)(T-T_c)}{\dot{m}_L c_{pL} + \dot{m}_G c_{pG}} \quad (6.78)$$

Tank reactor:

$$\frac{T-T_0}{V_R} = \frac{\int_0^{\delta_L} \sum_{j=1}^{s} R_j(-\Delta H_{rj})dza_v + \sum_{j=1}^{s} R_j(-\Delta H_{rj})\varepsilon_L - U\left(\frac{S}{V_R}\right)(T-T_c)}{\dot{m}_L c_{pL} + \dot{m}_G c_{pG}} \quad (6.79)$$

The integrals in the equations describe the contribution of the film to the balance equation, and it is obtained from the solution of the liquid film equations, as described in Section 6.2.

It should be remembered that the gas and liquid phases are presumed to have the same temperature in the above equations. The contributions of the gas and the liquid to the energy balance can be checked by calculating the terms in the denominators. It should be noticed that the gas and liquid phases can have different temperatures in column reactors; for such cases, separate balance equations are needed.

6.6 Physical properties of gas–liquid systems

The essential physical properties needed in gas–liquid reactor modelling are discussed briefly in this section. The most important quantities are the diffusion coefficients for the gas and liquid phases. The mass transfer coefficients appearing in the models are not discussed, since the topic is scattered and somewhat controversial: many correlations have been published in literature, but in connection with any new design project, it is necessary to conduct a new search in the most recent publications.

6.6.1 Diffusion coefficients in gas and liquid

6.6.1.1 Gas phase
The Fuller-Schettler-Giddings equation gives the binary diffusion coefficient in the gas phase (Reid et al. (1988)):

$$D_{ik} = \frac{(T/K)^{1.75} \left[\left(\frac{(g/mol)}{M_i}\right) + \left(\frac{(g/mol)}{M_k}\right) \right]^{\frac{1}{2}} \cdot 10^{-7}}{(P/atm)\left(v_i^{1/3} + v_k^{1/3}\right)^2} \quad (m^2/s) \tag{6.80}$$

These binary diffusion coefficients are used directly if the diffusion is described by Stefan-Maxwell equations. In case of simpler diffusion models, the single-component diffusion coefficients are then obtained from the Wilke approximation (Reid et al. (1988)):

$$D_{mi} = \frac{c - c_i}{\sum_{k=1, k \neq i}^{N} \frac{c_k}{D_{ik}}} \tag{6.81}$$

where c is the total concentration of gas obtained from the gas law, $c = \sum c_i$, and D_{ik} denotes the binary diffusion coefficient calculated from eq. (6.80). According to the film theory, we have the relations

$$D_i = D_{Gi} \quad \text{and} \quad k_{Gi} = \frac{D_{Gi}}{\delta_i} \tag{6.82}$$

It should be noted, however, that the mass transfer coefficient is not in practice proportional to the first power of the diffusion coefficient, but existing correlations typically predict a power in the range of ~0.6. The more advanced surface renewal theory of Danckwerts (1970) predicts the power of 0.5 highly realistically.

6.6.1.2 Liquid phase
The theoretical concept of liquid-phase diffusion is not very well developed. This is due to the complexity of the liquid phase, for example, aggregation of molecules, attraction and repulsion between ionic species and the complex character of molecular collisions in the liquid phase. Semi-empirical equations for liquid-phase diffusion coefficients use the Stokes-Einstein (S-E) formula as a starting point. According to S-E equation (Reid et al. (1988)), the diffusion coefficient of a solute (A) in a solvent (B) is obtained from

$$D_{AB} = \frac{RT}{6\pi\mu_B R_A} \tag{6.83}$$

where μ_B is the solvent viscosity and R_A is the gyration radius of the molecule A.

Several modifications of the S-E equation have been suggested in order to change the molecule radius into a more tractable, simpler quantity and to improve correspondence with experimental data. Wilke and Chang (1955) have proposed the formula:

$$D_{AB} = \frac{7.4 \times 10^{-12}\sqrt{[\Phi M_B(g/mol)]}(T/K)}{(\mu_B/cP)V_A^{0.6}} \quad (m^2/s) \tag{6.84}$$

where Φ is the association factor of the solvent, M_B is the solvent molar mass and V_A is the molar volume of the solute at the normal boiling point.

The association factor describes the aggregation of molecules; for instance, non-polar hydrocarbons do not aggregate at all, while water forms oligomers through hydrogen bonds. Thus, the most common solvent, water, is the most cumbersome one to be described. Values of some association factors (Φ) are listed in Table 6.3.

Table 6.3: Association factors for some common solvents.

Solvent	Φ
Water	2.6
Methanol	1.9
Ethanol	1.5
Non-polar organic solvents	1.0

Several modifications of the original Wilke-Chang formula exist; the reader is referred to Reid et al. (1988). The benefit of the Wilke-Chang equation is that it is easily extended to solvent mixtures, for which we propose:

$$D_{Am} = \frac{7.4 \cdot 10^{-12}\sqrt{\phi M}(T/K)}{\mu_m V_A^{0.6}} \quad (m^2/s), \quad \text{where} \quad \phi M = \sum_{i=1}^{N} x_i \phi_i M_i \tag{6.85}$$

It should be noted that eq. (6.85) contains solvent viscosity (μ_m), which is crucial for correct prediction of the diffusion coefficient. The association factor, ϕ_i, is determined empirically. V_A is the molar volume of the solved molecule at the normal boiling point. If experimental data for V_i:s is not available, it can be estimated from molar (atomic) increments of Le Bas (Reid et al. (1988)). Some atomic increments are listed in Table 6.4.

The Scheibel correlation (Scheibel (1954)) is a modification of the Wilke-Chang equation that eliminates the use of association factor and estimates the infinite diffusivities:

Table 6.4: Atomic increments of Le Bas for calculations of molar volumes.

Atom	Increment (cm³/mol)
Carbon	14.8
Hydrogen	3.7
Oxygen*	9–12
Chlorine	24.6

*Depending on the compound

$$D_{AB}^0 = \frac{8.2 \cdot 10^{-8} \cdot T}{\mu_B V_A^{1/3}} \left[1 + \left(\frac{3 V_B}{V_A}\right)^{2/3}\right] (cm^2/s) \tag{6.86}$$

where D_{AB}^0 is the diffusion coefficient for a dilute solute A into the solvent B.

Tyn and Calus (Reid et al. (1988)) have suggested a simple formula for the estimation of V_i

$$\left(\frac{V_i}{cm^3/mol}\right) = 0.285 \left(\frac{V_c}{cm^3/mol}\right)^{1.048} \tag{6.87}$$

where V_c is the critical volume of the solute. This value is accessible in various databases (e.g. Reid et al. (1988)).

The solvent viscosity, μ_B, is needed in the Wilke-Chang equation. Several empirical formulae for μ exist, such as

$$\mu = A T^B \tag{6.88a}$$

$$ln(\mu) = A + \frac{B}{T} + CT + DT^2 \tag{6.88b}$$

$$ln(\mu) = A + \frac{B}{T} \tag{6.88c}$$

These equations are of an empirical character; typically, the coefficients $(A \ldots D)$ are obtained from experimental viscosity data by regression analysis. Coefficient values for pure components are listed in ref. (Reid et al. (1988)).

For concentrated binary solutions, the Vignes equation (Vignes (1966)) can be used to predict the diffusion coefficient from the infinite dilute coefficients as a function of the composition

$$D_{1,2} = D_{2,1} = \left(D_{1,2}^0\right)^{x_2} \cdot \left(D_{2,1}^0\right)^{x_1} \tag{6.89}$$

For concentrated multicomponent systems, several mixing rules are available from the literature. A commonly used one is the Perkins and Geankoplis equation (Perkins and Geankoplis (1969)):

$$D_{Am}\mu_m^{0.8} = \sum_{\substack{i=1 \\ i \neq A}}^{n} x_i D_{A,i}^0 \mu_i^{0.8} \qquad (6.90)$$

where μ_m is the viscosity of the liquid mixture, μ_i is the viscosity of compound i.

For solvent mixtures, the case is more tricky. Theories exist for the prediction of viscosities of solvent mixtures, but the parameters needed in the formulae are usually missing. A general rule of thumb is that the viscosity of a solvent mixture should be determined experimentally. Experimental viscosity measurements are simple, inexpensive and reliable.

Basically, we only discussed just one equation for liquid viscosity above. It is, however, not at all guaranteed that the Wilke-Chang equation is the best choice. Very much depends on the solute and solvent molecules in the case. Particularly for aqueous solutions, special correlations have been developed for diffusion coefficients. An excellent and critical overview of various equations proposed for liquid-phase diffusion is provided by Wild and Charpentier (1987).

6.6.2 Gas–liquid equilibrium

The gas–liquid equixlibrium ratio used in previous derivations (Section 6.2) is defined by:

$$K_i = \frac{c_{Gi}^s}{c_{Li}^s} \qquad (6.91)$$

where the subscript (s) refers to interfacial concentrations. De facto, this equilibrium ratio is not a true constant as will be shown below. According to the laws of thermodynamics, the equilibrium ratio is defined using the mole fractions

$$T_i = \frac{y_i}{x_i} \qquad (6.92)$$

where x_i and y_i denote the mole fractions in liquid and gas. For concentrations of individual species, we have

$$c_G = \sum_i c_{Gi}, \quad c_L = \sum_i c_{Li} \qquad (6.93)$$

which are inserted in the equilibrium ratio

$$K_i = \frac{y_i}{x_i} \frac{c_G}{c_L} = K_{T_i} \frac{c_G}{c_L} \qquad (6.94)$$

This shows how the concentration-based equilibrium ratio is related to the thermodynamic quantity. The thermodynamic ratio is a function of composition, temperature and total pressure, that is $K_{T_i} = f(\underline{x}, \underline{y}, P, T)$. Thermodynamics tells us that this ratio is related to the activity (fugacity) coefficients of the gas (G) and liquid (L) phases:

$$\frac{y_i}{x_i} = K_{T_i} = \frac{\Phi_i^L}{\Phi_i^G} \qquad (6.95)$$

The activity coefficients for the gas phase are obtained in an equation of state, such as Soave-Redlich-Kwong or Peng-Robinson equations (Reid et al. (1988)), while the activity coefficients are usually calculated from specific theories for liquid phase, such as UNIQUAC-UNIFAC (Reid et al. (1988)). The reader is referred to special literature on the topic.

The complete, rigorous thermodynamic modelling of gas–liquid equilibria described above is frequently used in systems where all of the components principally coexist in gas and liquid phases. These cases are very common in distillation processes, but not that common in gas–liquid reactions and absorption processes. Typically, few components exist in the gas phase, and the volatility of most liquid-phase components is negligible. For these kinds of limited gas solubility, the simpler relation proposed by Henry can be used: Henry's law states that gas solubility depends exclusively on temperature and, thus, the constant in eq. (6.95) is a "true constant." Usually Henry's law is given in the form,

$$He = \frac{p_i}{c_{Li}} \text{ or } He' = \frac{p_i}{x_{Li}} \qquad (6.96)$$

Tabulated data for Henry's constants as a function of temperature are presented in the comprehensive treatment of Fogg and Gerrard (1991). Empirically, the temperature dependence of gas solubility can be described with an equation of the following kind:

$$\ln(x) = A + \frac{B}{T} + C \cdot \ln(T) + D \qquad (6.97)$$

where A, B, C and D are empirical coefficients, listed in Table 6.5.

6.7 Numerical strategies for gas–liquid reactor models

Mathematical structures of the gas–liquid reactor models are summarized in Table 6.6. Since film equations are included, numerical methods for differential equations, boundary value problems, are needed. For column reactor models, the derivatives with respect to the column length imply that partial differential equations are involved. The key point is how to discretize the film and column equations. A characteristic feature is the existence of very steep profiles in the liquid films, particularly for rapid reactions, in which all gaseous reactants are consumed in the liquid film. This implies that global approximation methods, such as orthogonal collocation with a single polynomial, are

Table 6.5: Selected gas solubilities according to eq. (6.95). Data from Fogg and Gerrard (1991).

Gas	Solvent	A	B	C	D	Temperature interval/K
H_2	H_2O	−123.939	5528.45	16.8893	0	273–345
H_2	Hexane	−5.8952	−424.55	0	0	213–298
	Heptane	−5.6689	−480.99	0	0	238–308
	Octane	−5.6624	−484.38	0	0	248–308
	Benzene	−5.5284	−813.90	0	0	280–336
	Toluene	−6.0373	−603.07	0	0	258–308
	Methanol	−7.3644	−408.38	0	0	213–298
	Ethanol	−7.0155	−439.18	0	0	213–333
O_2	H_2O	−171.2542	8391.24	23.24323	0	273–617
	H_2O	−139.458	6889.6	18.554	0	283–343
	Benzene	−30.1649	874.16	3.53024	0	248–343
	Ethanol	−7.874	126.93	0	0	273–353
CO_2	H_2O	−159.854	8741.68	21.6694	$-1.10261 \cdot 10^{-3}$	283–313
	Benzene	−73.824	3804.8	9.8929	0	283–313
	Toluene	−13.391	1512.9	0.6580	0	203–313

Table 6.6: Mathematical structures of gas–liquid reactor models.

	Tank reactor model
Dynamic	ODEs + parabolic PDEs
Steady state	NLEs + ODEs (BVP)
	Column reactor model (with axial dispersion)
Dynamic	Parabolic PDEs
Steady state, co- or countercurrent	ODEs BVP

not particularly suitable for film equations. Spline collocation, that is, collocation on finite elements, works well: the film is divided into subintervals, where orthogonal collocation is applied to individual intervals and the continuity of the first derivative is required (Romanainen and Salmi (1995)).

Alternatively, finite differences can be applied to the discretization. The use of equidistant discretization requires a good number of discretization points but, as a compensation, higher order difference formulae can be applied (see Table 6.7). Romanainen and Salmi (1995) have proposed the use of a half-discretization scheme illustrated in Figure 6.7. The benefit is the accumulation of points close to the phase boundary. The method is suitable for extremely steep profiles. In practise, it is necessary to apply the half-even discretization procedure illustrated in Figure 6.7: in the

Table 6.7: Finite difference formulae.

	Coefficients A_{mk} for the multipoint central differentiation of the second derivative								
m	A_{m0}	A_{m1}	A_{m2}	A_{m3}	A_{m4}	A_{m5}	A_{m6}	A_{m7}	A_{m8}
$3^\dagger$	$\frac{1}{1}$	$-\frac{2}{1}$	$\frac{1}{1}$						
$5^\dagger$	$-\frac{1}{12}$	$\frac{16}{12}$	$-\frac{30}{12}$	$\frac{16}{12}$	$-\frac{1}{12}$				
$7^\ddagger$	$\frac{2}{180}$	$-\frac{27}{180}$	$\frac{270}{180}$	$-\frac{490}{180}$	$\frac{270}{180}$	$-\frac{27}{180}$	$\frac{2}{180}$		
$9^\ddagger$	$-\frac{9}{5040}$	$\frac{128}{5040}$	$-\frac{1008}{5040}$	$\frac{8064}{5040}$	$-\frac{14350}{5040}$	$\frac{8064}{5040}$	$-\frac{1008}{5040}$	$\frac{128}{5040}$	$-\frac{9}{5040}$

To illustrate

3-point formula: $\left(\frac{d^2y}{dx^2}\right)_{x_0} = \frac{\left(\frac{1}{1}y_{-1} - \frac{2}{1}y_0 + \frac{1}{1}y_{+1}\right)}{\Delta x^2}$

	Coefficients B_{mk} for the multipoint backward differentiation of first derivative							
M	B_{m0}	B_{m1}	B_{m2}	B_{m3}	B_{m4}	B_{m5}	B_{m6}	B_{m7}
$1^\dagger$	$-\frac{1}{1}$	$\frac{1}{1}$						
$2^\dagger$	$\frac{1}{2}$	$-\frac{4}{2}$	$\frac{3}{2}$					
$3^\dagger$	$-\frac{2}{6}$	$\frac{9}{6}$	$-\frac{18}{6}$	$\frac{11}{6}$				
$4^\dagger$	$\frac{3}{12}$	$-\frac{16}{12}$	$\frac{36}{12}$	$-\frac{48}{12}$	$\frac{25}{12}$			
$5^\dagger$	$-\frac{12}{60}$	$\frac{75}{60}$	$-\frac{200}{60}$	$\frac{300}{60}$	$-\frac{300}{60}$	$\frac{137}{60}$		
$6^\ddagger$	$\frac{10}{60}$	$-\frac{72}{60}$	$\frac{225}{60}$	$-\frac{400}{60}$	$\frac{450}{60}$	$-\frac{360}{60}$	$\frac{147}{60}$	
$7^\ddagger$	$-\frac{60}{420}$	$\frac{490}{420}$	$-\frac{1764}{420}$	$\frac{3675}{420}$	$-\frac{4900}{420}$	$\frac{4410}{420}$	$-\frac{2940}{420}$	$\frac{1089}{420}$

To illustrate

2-point formula: $\left(\frac{dy}{dx}\right)_{x_0} = \frac{\left(\frac{1}{1}y_0 - \frac{1}{1}y_{-1}\right)}{\Delta x}$

3-point formula: $\left(\frac{dy}{dx}\right)_{x_0} = \frac{\left(\frac{3}{2}y_0 - \frac{4}{2}y_{-1} + \frac{1}{2}y_{-2}\right)}{\Delta x}$

† from Abramowitz and Stegun (1970)
‡ derived during this work

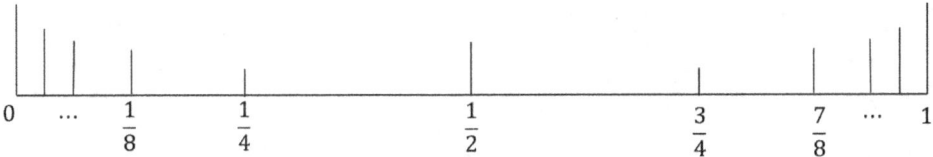

Figure 6.7: Half-even discretization procedure.

central part of the film, an even discretization scheme is used, and it is replaced by a half-discretization scheme in the vicinity of the phase boundary.

Romanainen and Salmi (1995) made an extensive comparison of various numerical approaches for gas–liquid reactor models (Table 6.8). Generally, we can conclude that spline collocation applied to the films and column coordinates gives a very accurate solution. In general, the solution of balances in the dynamic form with respect to time is a very robust approach (Figure 6.8).

Table 6.8: Overview of numerical strategies for gas–liquid reactors.

Gas–liquid tank reactor, steady state	Discretize the liquid film with finite differences or orthogonal collocation, solve the model as a set of NLE (include the film and bulk phase balances) (Appendix A.1)
Gas–liquid reactor, dynamic	Film discretization as above, solve the set of ODEs created with a stiff ODE solver (Appendix A.2)
Gas–liquid column reactor, steady state	Double discretization of the film and reactor length coordinate, solve the set of NLEs created by an NLE solver (Appendix A.1)
Gas–liquid column reactor, dynamic	Double discretization of the space coordinates as above, solve the ODEs created with a stiff ODE solver (Appendix A.2)

A final remark is worth noting: before launching the heavy numerical machinery, it is wise to investigate carefully the rapidity of the reaction compared to the speed of diffusion, that is, by evaluating the Hatta number (Table 6.2). For low Hatta numbers, the complete reaction-diffusion model for the liquid film is not needed, but we can keep to a simplified treatment, accounting for diffusion alone (no reaction) in the liquid film. For the lowest Hatta numbers, even a saturation state of the liquid phase can be assumed, and the reactor model de facto becomes similar to the models of homogeneous liquid-phase reactors (Chapter 3).

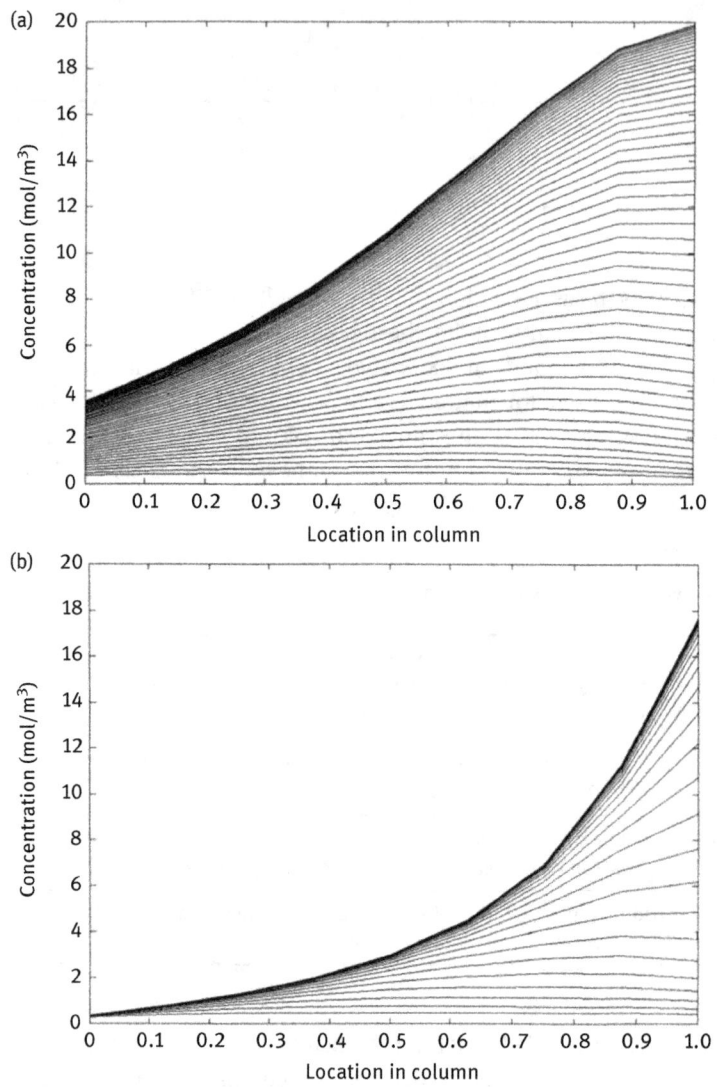

Figure 6.8: Simulation of bubble columns. Case study: Chlorination of p-cresol (Romanainen J., Numerical approach to modeling of dynamic bubble columns, 1997).

7 Equipment and models for laboratory experiments

The reader might well ask why a separate chapter devoted to laboratory reactors is needed. From a purely theoretical viewpoint, it is unnecessary: all laboratory reactors can be placed in the well-known categories of homogeneous and heterogeneous reactors. However, laboratory reactors typically possess some characteristic features which are worth discussing: the residence time distribution and temperature are carefully controlled, and the flow pattern is maintained as simple as possible. This is due to the primary purpose of laboratory-scale reactors: they are mainly used to screen and determine the kinetics and equilibria of chemical processes, preferably in the absence of heat and mass transfer limitations. Laboratory reactors have a special design to suppress the above-mentioned effects, and automated data acquisition is used. The specific phenomena attributed to laboratory reactors will be discussed in this chapter. Primary data obtained from laboratory reactors is used to determine parameter values important for process design. For further data processing and determination of parameter values, it is important to recognize the mathematical structure of the model. This is a further topic of the present Chapter. Processing of experimental data by numerical analysis will be discussed in detail in Chapter 8.

7.1 Homogeneous batch reactor

The characteristic feature of a homogeneous batch reactor is that the volume is constant for gas-phase processes, and in practice often also for liquid-phase processes. Concentrations in the gas or liquid phase are measured as a function of reaction time. Various methods for chemical analysis are nowadays accessible; continuous methods applied on line, such as conductometry and photometry, on-line infrared as well as discontinuous methods based on sampling and off-line analysis of the samples, such as gas and liquid chromatography. A typical kinetic experiment is shown in Figure 7.1. Provided that stirring is vigorous in the reactor, the mass balance is written as:

$$\frac{dc_i}{dt} = r_i(\underline{c}, \underline{k}, \underline{K}) \qquad (7.1)$$

De facto, eq. (7.1) is a special case of the equations presented in Chapter 3 for homogeneous tank reactors. In the evaluation of kinetic data displayed in Figure 7.1, two methods are used, the differential method and the integral method. These methods are treated in detail in numerous textbooks devoted to chemical kinetics and chemical reaction engineering. A simple example will be given below.

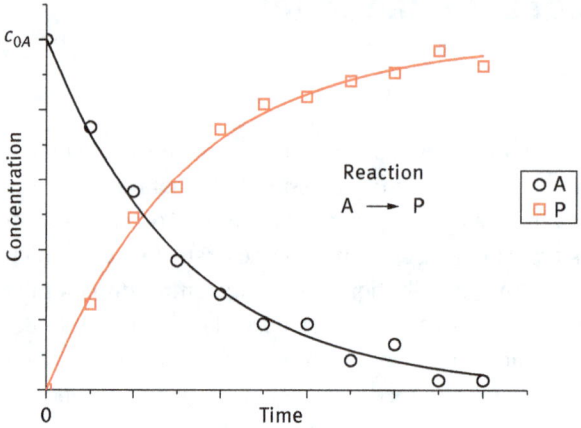

Figure 7.1: A typical kinetic experiment in a batch reactor.

Let us assume that a batch reactor experiment for the reaction $A \rightarrow P$ has been conducted in the liquid phase. The concentration of A, c_A, has been measured as a function of time, and a first-order rate law is assumed; $r = kc_A$. Thus, the generation rate of A becomes

$$r_A = -r = -kc_A \tag{7.2}$$

After inserting this expression in the mass balance (7.1), we obtain the differential equation

$$-\frac{dc_A}{dt} = kc_A \tag{7.3}$$

Firstly, the **differential method** is used in the following way to obtain the value of the rate constant:
1. Obtain c_A as a function of t, $c_A(t)$ experimentally
2. Estimate dc_A/dt at each point $(t, c_A(t))$ by numerical differentiation
3. Plot $-dc_A/dt$ as a function of c_A. The slope of the curve gives the rate constant.

This procedure is illustrated in Figure 7.2.

In practise, step 3 is performed using linear regression analysis. Numerous computer software packages contain routines for linear regression of the type $y = ax$ and $y = ax + b$.

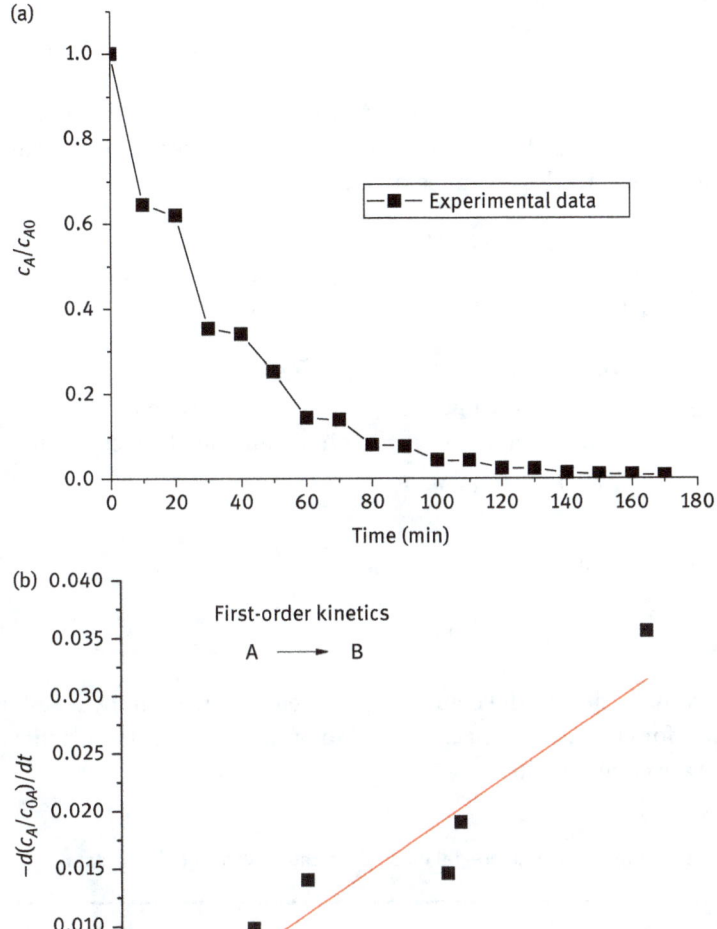

Figure 7.2: Typical kinetic data from a batch reactor: analysis by differential method.

The **integral method** proceeds as follows:
1. Integrate the balance equation analytically or numerically; for our simple example, we obtain an easy separation of variables:

$$-\int_{c_{A0}}^{c_A} \frac{dc_A}{c_A} = k \int_0^t dt \qquad (7.4)$$

$$-\ln\left(\frac{c_A}{c_{A0}}\right) = kt \tag{7.5}$$

It can be shown that the procedure can be generalized to apply to many kinds of kinetics. Generally, a concentration-dependent function and a time-dependent function are obtainable after separation of variables and integration:

$$F(\underline{c_A}) = kt \tag{7.6}$$

2. Plot $F(\underline{c_A})$ (here $-\ln(c_A/c_{A0})$) as a function of time, t. The slope is equal to k.

Step 2 is performed by regression analysis. Again, we have a linear problem with respect to the parameter (k), and existing linear regression software can be applied.

In a very general case, it is not even necessary to obtain an analytical solution

$$F(\underline{c_A}) = kt \tag{7.7}$$

of the original ODE, but the balance

$$\frac{dc_A}{dt} = r_A \tag{7.8}$$

can be solved numerically in situ, in the course of parameter estimation, by a suitable numerical method for ODEs, as discussed in Chapter 3. De facto, the benefits of the two approaches are compared in Table 7.1.

Table 7.1: A comparison of the differential and integral methods in the analysis of kinetic data.

Differential	Integral
+ Principally simple	− Non-linear regression often needed
+ Linear regression sometimes possible	+ The principally correct method
− Numerical differentiation is tricky; special tricks to maintain accuracy	+ Superior to differential method in accuracy

As Table 7.1 shows, the integral method is preferable in the analysis of batch reactor data, as the risky step of numerical differentiation is avoided. Generally, it can be stated that the differential method is useful in a preliminary screening of data, while the final analysis should be carried out by means of the integral method.

7.2 Homogeneous stirred tank reactor (CSTR)

For a homogeneous CSTR, constant volume and pressure are reasonable assumptions. Concentrations at the reactor outlet are measured as a function of the space time, that is, the volumetric flow rate. The steady-state mass balance is written as (Chapter 3),

$$\frac{c_i - c_{0i}}{\tau} = r_i(\underline{c}, \underline{k}, \underline{K}) \tag{7.9}$$

where

$$\tau = \frac{V}{\dot{V}_0}, \qquad \dot{V} = \dot{V}_0 \tag{7.10}$$

It should be noted that the volumetric flow rates vary considerably for gas-phase reactions, as discussed in Chapter 3. Measurements of c_i (and c_{0i}, τ) give directly r_i, that is, the generation rates of the compounds, which is the great benefit of this reactor as a rapid tool for kinetic experiments. With a CSTR, very short residence times become possible, which implies that the kinetics of rapid reactions can be measured. On the other hand, a large quantity of chemicals is consumed during an experiment.

An example of the use of a CSTR is given below. The reaction $A \to P$ is studied in a CSTR. The space time, τ – and possibly the initial concentration level, c_{0A} – are varied and c_A, r_A data are obtained. A rate law is assumed, for example, the reaction rate for the component A is $r_A = -kc_A$ according to first-order kinetics. A concentration plot of the component A, c_A, versus $(-r_A)$ defines the reaction rate constant, k, as the slope of the curve, as illustrated in Figure 7.3.

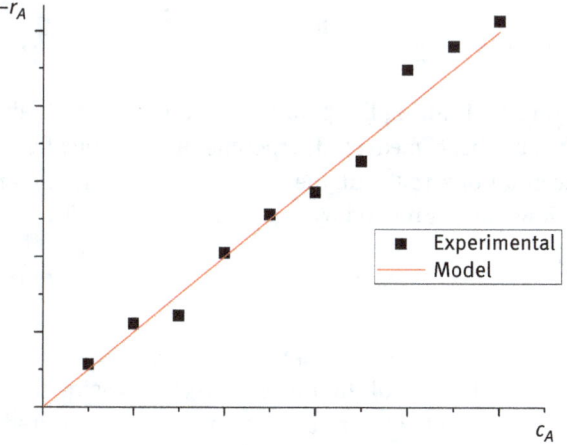

Figure 7.3: Determination of rate constant from data obtained from a CSTR.

Linear regression is used to obtain the rate constant, k, from the data. The crucial point in the use of a CSTR as a test reactor is to measure the space time ($\tau = V_R/\dot{V}_0$) very precisely, since its value (τ) affects directly all other data. Furthermore, steady-state conditions need to prevail with certainty.

7.3 Catalytic fixed bed in integral mode

Catalytic fixed beds are frequently used as test reactors for two-phase processes (gas or liquid and a solid catalyst). To this category of reactors belong conventional fixed beds but also structured reactors, such as monoliths and microreactors, provided that the assumption of a plug flow is valid. Plug flow conditions should prevail and small enough catalyst particles should be used to suppress the diffusion resistance (Chapters 4–5). Provided that these conditions are fulfilled, the mass balance becomes very simple:

$$\frac{d\dot{n}_i}{dV} = \rho_B r_i \tag{7.11}$$

Equation (7.11) is de facto similar to the one-dimensional, pseudo-homogeneous model presented in Chapter 4. The mass of the catalyst in the reactor is m_{cat},

$$m_{cat} = \rho_B V_R \tag{7.12}$$

and we introduce a dimensionless coordinate, accordingly

$$V = V_R z, \qquad dV = V_R dz \tag{7.13}$$

These give a new form of the balance equation

$$\frac{d\dot{n}_i}{dz} = m_{cat} r_i, \qquad c_i = \frac{\dot{n}_i}{\dot{V}} \tag{7.14}$$

Equation (7.14) is solved from $z = 0$ to $z = 1$ during the parameter estimation to obtain the molar flows, $\dot{n}_i$ ($z = 1$), which have been measured experimentally. Typically, a chemical analysis gives the concentrations (c_i), but they are related to the molar flows by $\dot{n}_i = c_i \dot{V}_i$. The volumetric flow rate is updated by a gas law:

$$\dot{V} = \frac{Z\dot{n}RT}{p}, \qquad \dot{n} = \sum_i \dot{n}_i \tag{7.15}$$

The inlet composition is varied in the experiments to get $\dot{n}_i$ under different conditions.

For simple cases of kinetics, an analytical solution of eq. (7.14) is possible but, in general, a numerical solution in the course of parameter estimation is preferred. A special case of the fixed bed reactor model is considered in the next section.

7.4 Catalytic differential reactor

Catalytic differential reactors are frequently mentioned in textbooks as typical test reactors. In fact, a catalytic differential reactor is a special case of the fixed bed reactor, nothing else. Conversions are kept low, which allows us to approximate the molar flow gradients by a linear function

$$\frac{d\dot{n}_i}{dz} \approx constant = \frac{\dot{n}_i - \dot{n}_{i0}}{1-0} = \dot{n}_i - \dot{n}_{i0} \qquad (7.16)$$

The approximation is illustrated in Figure 7.4.

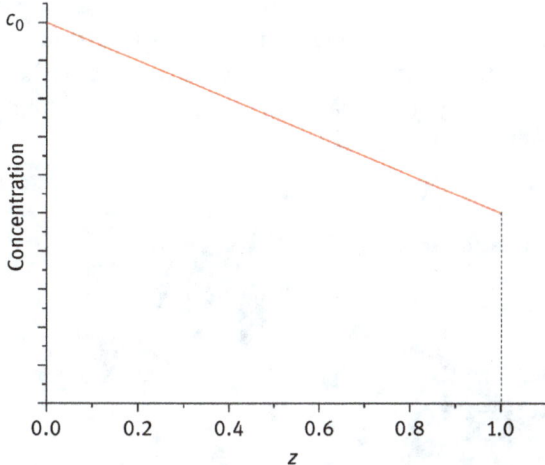

Figure 7.4: Concentration gradients in a differential reactor.

As the Figure shows, the concentration gradients are practically linear as long as the conversion is low, that is, the rate constant is low and/or the space time is short. By denoting

$$\dot{n}_i = \dot{n}_i(z=1) \qquad (7.17)$$

and inserting this relation in the balance equation of a catalytic fixed bed, eq. (7.14), we obtain

$$\dot{n}_i - \dot{n}_{i0} = m_{cat}\overline{r}_i \qquad (7.18)$$

where $\overline{r}_i$ indicates the generation rate, which should be calculated as an average of the concentrations or molar flows.

$$\overline{\dot{n}}_i = \frac{1}{2}(\dot{n}_{i0} + \dot{n}_i) \qquad (7.19)$$

The balance can, thus, be rewritten as

$$\overline{\dot{n}_i} - \dot{n}_{i0} = \frac{1}{2} m_{cat} \overline{r_i} \qquad (7.20)$$

In principle, $\overline{r_i}$ is directly obtained by means of a measurement of the molar flow difference, $\overline{\dot{n}_i} - \dot{n}_{i0}$. In a mathematical sense, the differential reactor model coincides with the model of a gradientless catalytic test reactor (CSTR) presented in the following section.

7.5 Catalytic gradientless reactor

In a catalytic gradientless reactor, a constant volume and pressure is assumed. Complete backmixing is achieved by different means, such as by a rotating basket (spinning basket) inside the reactor, shaking of the reactor contents or by recycling. Different configurations are illustrated in Figure 7.5.

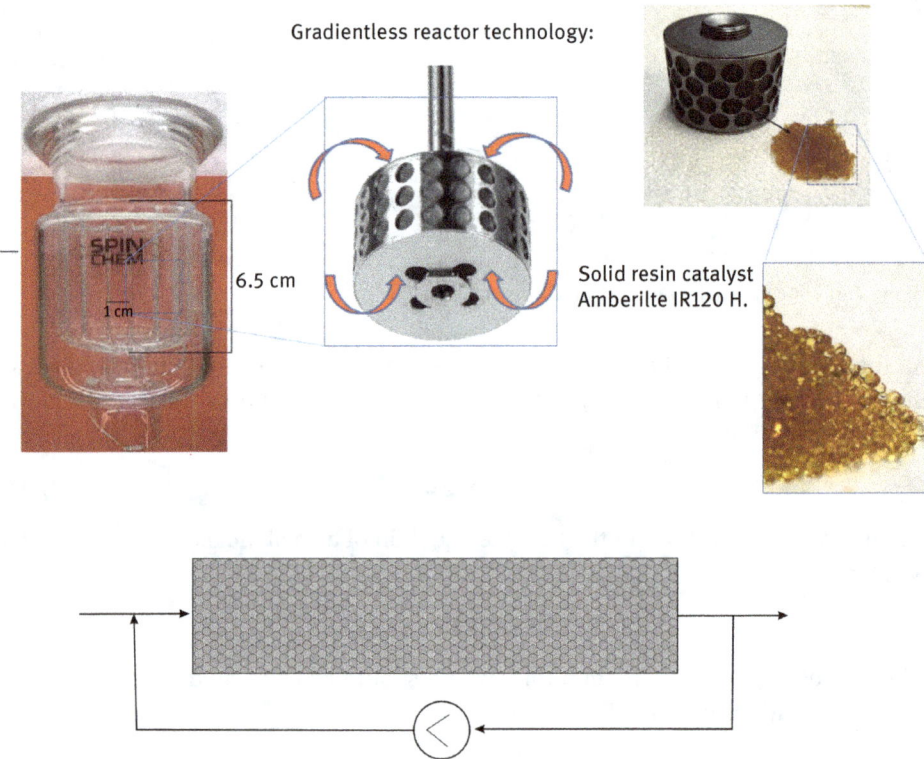

Figure 7.5: Configuration alternatives for a gradientless reactor system (SpinChem) and a recycle reactor.

In a steady state, the mass balance or an arbitrary component, according to the principles presented in Chapter 4, becomes:

$$\dot{n}_i - \dot{n}_{i0} = m_{cat} r_i \tag{7.21}$$

Thus, the generation rates are obtained directly by measuring $\dot{n}_i$ or c_i at the reactor outlet.

7.6 Catalytic slurry reactor

For catalytic three-phase processes, batchwise operating slurry reactors are frequently used for kinetic experiments. The gas–liquid and liquid–solid mass transfer resistances are suppressed by vigorous stirring, and the mass-transfer resistance inside the catalyst particles is minimized by using finely dispersed particles on a micrometre scale (Chapter 5). The addition of the gas-phase component is controlled by pressure regulation; thus pressure in the gas-phase is kept constant, which implies that the mass balance of the gas-phase can be ignored in the mathematical treatment. The concentrations of dissolved gases in the liquid phase are equal to the saturation concentrations (Chapter 5). Under these circumstances, the mass balance of an arbitrary non-volatile component in the liquid phase becomes

$$\frac{dc_i}{dt} = \rho_B r_i, \qquad \text{where } \rho_B = \frac{m_{cat}}{V_L} \tag{7.22}$$

As we can see, the formal treatment of the data coincides with the treatment of data from homogeneous batch reactors. The only difference arises from using the proportionality factor, that is, the catalyst bulk density, ρ_B. Some analytical solutions of eq. (7.22) were presented in Chapter 5.

7.7 Classification of laboratory reactor models

Models that are relevant in the estimation of kinetic and thermodynamic parameters are reviewed in this section. Roughly speaking, two kinds of models are highly dominating, namely algebraic models and differential models. Whether the model is algebraic or differential depends on the reactor and its state (steady state or dynamic state).

7.7.1 Algebraic and differential models

Algebraic models can be represented by the equation

$$\underline{\hat{y}} = f(\underline{x}, \underline{p}) \tag{7.23}$$

where $\hat{y}$ is the dependent variable representing, for instance, the rates measured in CSTRs. Equally well, the dependent variable can represent the measured rates in differential reactors or analytically integrated balances of batch reactors. Examples of algebraic models will be given in the next section.

Differential models can generally be described by ordinary differential equations (ODEs), such as

$$\frac{d\underline{\hat{y}}}{d\theta} = f(\underline{x}, \underline{p}) \tag{7.24}$$

where the symbol θ represents time, length, volume etc., depending on the particular case. Partial differential equations, particularly hyperbolic and parabolic ones, can be transformed into the form presented in eq. (7.24) by discretization, as discussed in previous Chapters.

The dependent variable ($\hat{y}$) corresponds with the concentrations or molar amounts measured in batch and semi-batch reactors, or concentrations or molar flows at the outlet of plug flow, or, alternatively, fixed bed reactors.

7.7.2 Linearity and non-linearity of the model

By linearity and non-linearity of the model we mean linearity and non-linearity with respect to the model parameters, since they are the unknown quantities in the determination of kinetic and thermodynamic information. The following notations will be used here:

$\underline{Y}$	estimated dependent (measured) variable
$\underline{X}$	independent variable
$\underline{p}$	parameter

The concept of linearity-non-linearity is illustrated through a simple example: Assume that batch reactor experiments with a first-order reaction, $A \rightarrow P$, $r_A = -kc_A$, have been carried out. The concentration for component A, c_A, has been measured as a function of the reaction time.

The differential model for this system is given as the mass balance for a batch reactor:

$$\frac{dc_A}{dt} = -kc_A \qquad (7.25)$$

Translation of this model equation into the nomenclature used in parameter estimation implies that

$$\theta = t, \underline{y} = c_A, \underline{p} = k = p_1 \qquad (7.26)$$

We thus obtain the standard form of the differential model,

$$\frac{d\hat{\underline{y}}}{dt} = f(\underline{y}, \underline{p}) \qquad (7.27)$$

that is, in the present case

$$\frac{d\hat{\underline{y}}}{d\theta} = p_1 y \qquad (7.28)$$

where the reaction time, θ, is the independent variable, the concentration, $\hat{y}$, is the dependent variable and the rate constant, p_1, is the unknown parameter to be estimated by regression analysis.

It should be noted that the simple problem presented above can be solved as an algebraic model, too. Integration of the differential equation gives:

$$\int_{c_{A0}}^{c_A} \frac{dc_A}{c_A} = -k \int_0^t dt \qquad (7.29)$$

$$-\ln\left(\frac{c_A}{c_{A0}}\right) = +kt \qquad (7.30)$$

and we can define the variables and parameters for the regression analysis as follows

$$\underline{y} = -\ln\left(\frac{c_A}{c_{A0}}\right) = y_1, p_1 = k, \underline{x} = x_1 = t \qquad (7.31)$$

and obtain the simple and beautiful expression of the algebraic model: $y_1 = p_1 x_1$. This equation represents a linear algebraic model, since it is linear with respect to the parameter ($p_1 = k$).

On the other hand, if we use the form

$$\ln\left(\frac{c_A}{c_{A0}}\right) = -kt \qquad (7.32)$$

and solve the concentration in it, the result is

$$\frac{c_A}{c_{A0}} = e^{-kt} \qquad (7.33)$$

and the independent and dependent variables are defined along with the parameter,

$$y_1 = \frac{c_A}{c_{A0}}, \; p_1 = k, \; t = x_1 \qquad (7.34)$$

The model now becomes

$$y_1 = e^{-p_1 x_1} \qquad (7.35)$$

This is an algebraic model, too, but a non-linear one with respect to the parameter (p_1).

The use of algebraic and differential models will be illustrated in the following Chapter, as the concepts of non-linear regression analysis are presented.

8 Parameter estimation in reaction engineering

Models in chemical reaction engineering usually contain a number of unknown parameters whose values should be determined from experimental data. Regression analysis is a powerful and objective tool in the estimation of parameter values.

The task of regression analysis can be stated as follows: the values of the dependent variable (y) are predicted by the model; a function (f), contains independent variables (x) and parameters (p). The dependent variable is measured experimentally, in different conditions, that is, at different values of the independent variables (x). The goal is to find such numerical values of the parameters (p) that the model gives the best possible agreement with experimental data (Figure 8.1). Typical independent variables in chemical reaction engineering are reaction times, concentrations, pressures and temperatures, while molar amounts, concentrations, molar flows and reaction rates are dependent variables. The parameters to be estimated are usually rate and equilibrium constants, sometimes even mass and heat transfer coefficients. Since most models in chemical reaction engineering are non-linear with respect to the parameters, the discussion here is limited to non-linear regression analysis exclusively.

8.1 Principles of non-linear regression analysis

The model equation can be written as follows:

$$\underline{\hat{y}} = \underline{f}(\underline{x}, \underline{p}) \tag{8.1}$$

Strictly speaking, this is valid for algebraic models, but it can be applied to differential models as well. For differential models, the solution (y) is obtained numerically from the model equation as discussed in the preceding Chapters. The objective function (Q) to be minimized by non-linear regression is defined by:

$$Q(f) = \sum_{i=1}^{nm} \left[\underline{y_i} - \underline{f}(\underline{x_i}, \underline{p})\right]^T \underline{w_i} \left[\underline{y_i} - \underline{f}(\underline{x_i}, \underline{p})\right] \tag{8.2}$$

where $\underline{w_i}$ denotes the weight matrix of the experimental points.

After carrying out the multiplication, we get an easily understandable expression for the objective function,

$$Q(\underline{p}) = \sum_{i=1}^{nm} \left[(y_{1i} - f_{1i})^2 w_{11i} + (y_{2i} - f_{2i})^2 w_{22i} + \ldots (y_{nyi} - f_{nyi})^2 w_{ny,nyi}\right] \tag{8.3}$$

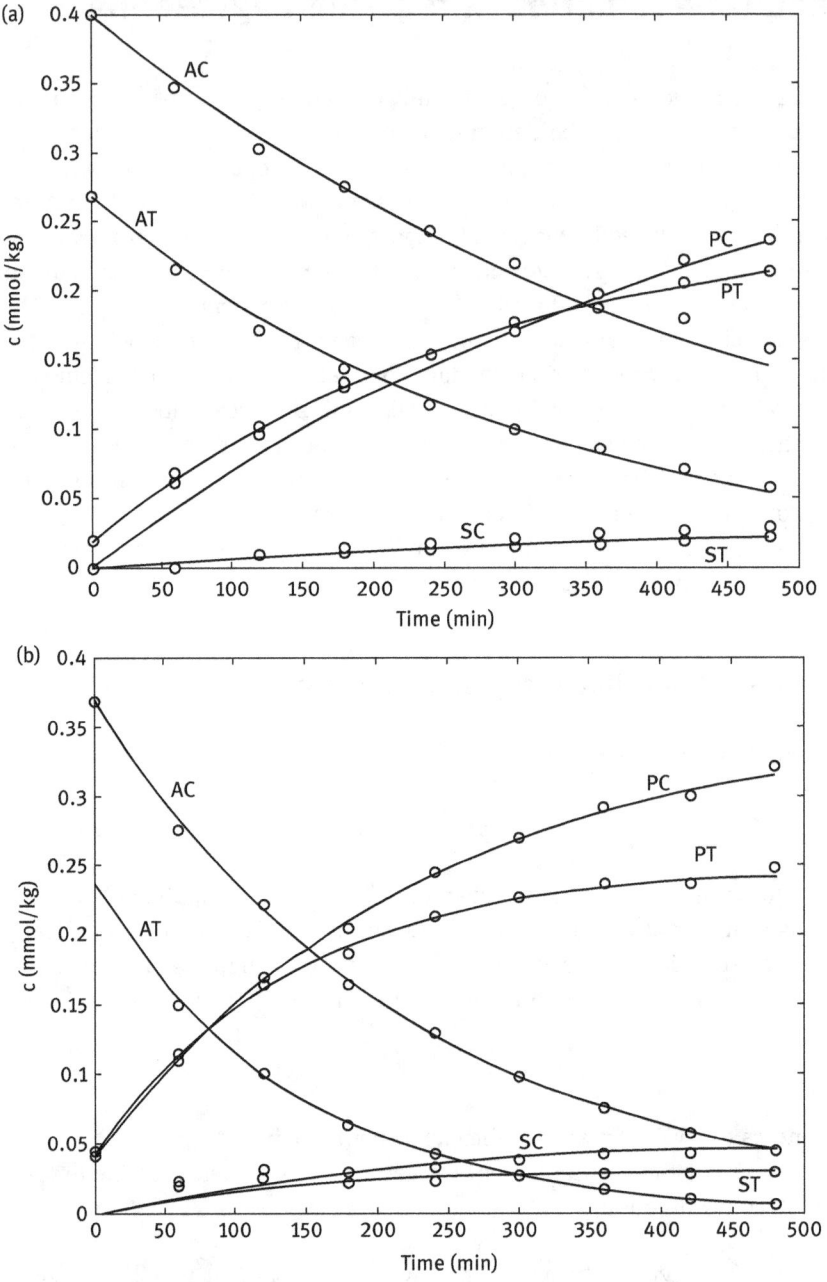

Figure 8.1: The goal of parameter estimation: a good description of experimental data (Grénman et al. (2003)).

8.1 Principles of non-linear regression analysis

In case we have a single independent variable and equal weight factors, the objective function is given by

$$N_y = 1, \quad w_i = 1, \quad i = 1 \ldots nm$$

$$w_i = [1\ 1\ 1 \ldots 1]$$

$$\underline{y} = (y_1\ y_2 \ldots y_{ny})^T$$

$$\underline{x} = (x_1\ x_2 \ldots x_{nx})^T$$

$$\underline{f} = (p_1\ p_2 \ldots p_{np})^T$$

$$Q(\underline{p}) = \sum_{i=1}^{nm} \left[(y_{1i} - f_{1i})^2 + (y_{2i} - f_{2i})^2 + \ldots (y_{nyi} - f_{nyi})^2 \right]$$

$$\begin{bmatrix} w_{11i} & 0 & \cdots & 0 \\ 0 & w_{22i} & \cdots & 0 \\ \vdots & \vdots & \ddots & \vdots \\ 0 & 0 & \cdots & w_{nyi} \end{bmatrix} \begin{bmatrix} y_{1i} - f_{1i} \\ y_{2i} - f_{2i} \\ \vdots \\ y_{nyi} - f_{nyi} \end{bmatrix}$$

In case the weight factors are equal, the objective function, eq. (8.3), is simplified to

$$Q(\underline{p}) = \sum_{i=1}^{nm} (y_i - f_i)^2 \qquad (8.4)$$

The necessary condition for obtaining a minimum value of the objective function, Q, is that all the partial derivatives with respect to the parameters are zero. Differentiation finally gives eq. (8.5), which is algebraic, and the function g is a system of np equations, the roots of which are found numerically,

$$\frac{\partial Q}{\partial p_j} = \frac{\partial}{\partial p_j} \left(\sum_{i=1}^{nm} (y_i - f_i)^2 \right) = 0, \quad j = 1 \ldots n_p \qquad (8.5)$$

$$\sum_i + 2(y_i - f_i) \frac{\partial f_j}{\partial p_j} = 0 = 2g_j \qquad (8.6)$$

$$\underline{g} = 0 \qquad (8.7)$$

The algebraic equations can be solved by the Newton-Raphson method, which gives the following algorithm for multiple equations (Appendix A.1):

$$\underline{p}_{m+1} = \underline{p}_m - G_m^{-1} \underline{g}_m \qquad (8.8)$$

where m denotes the iteration index. The Jacobian matrix consists of the elements

$$\underline{G} = \begin{bmatrix} \partial g_1/\partial p_1 & \partial g_1/\partial p_2 & \cdots & \partial g_1/\partial p_{np} \\ \partial g_2/\partial p_1 & \partial g_2/\partial p_2 & \cdots & \partial g_2/\partial p_{np} \\ \vdots & \vdots & \ddots & \vdots \\ \partial g_{np}/\partial p_1 & \partial g_2/\partial p_2 & \cdots & \partial g_{np}/\partial p_{np} \end{bmatrix} \qquad (8.9)$$

The elements of matrix ($\underline{G}$) are further elaborated,

$$\frac{\partial g_j}{\partial p_k} = \frac{\partial}{\partial p_k}\left[\sum_i \left(y_i - f_i \frac{\partial f_i}{\partial p_k}\right)\right] \qquad (8.10)$$

Equation (8.10) implies that the cross-derivatives are ignored.

$$\sum_i \left[-\frac{\partial f_i}{\partial p_k}\frac{\partial f_i}{\partial p_j} + (y_i - f_i)\frac{\partial^2 f_i}{\partial p_k \partial p_j} \right] \qquad (8.11)$$

$$\frac{\partial g_j}{\partial p_k} \approx -\sum_i \frac{\partial f_i}{\partial p_k}\frac{\partial f_i}{\partial p_j} \qquad (8.12)$$

By using the matrix (A),

$$\underline{A} = \begin{bmatrix} \partial f_1/\partial p_1 & \partial f_1/\partial p_2 & \cdots & \partial f_1/\partial p_{np} \\ \partial f_2/\partial p_1 & \partial f_2/\partial p_2 & \cdots & \partial f_2/\partial p_{np} \\ \vdots & \vdots & \ddots & \vdots \\ \partial f_{np}/\partial p_1 & \partial f_2/\partial p_2 & \cdots & \partial f_{np}/\partial p_{np} \end{bmatrix} \qquad (8.13)$$

the problem can be written in a very concise form:

$$-\underline{A}^T \cdot \underline{A} = \underline{G} \qquad (8.14)$$

The vector $\underline{f}$ is defined by

$$\underline{f} = \begin{bmatrix} y_{1i} - f_{1i} \\ y_{2i} - f_{2i} \\ \vdots \\ y_{nyi} - f_{nyi} \end{bmatrix} \qquad (8.15)$$

$$\sum_i (y_i - f_i)\frac{\partial f_i}{\partial p_j} = g_j \qquad (8.16)$$

Recalling the definition of the objective function, g, we arrive at eq. (8.17), which in terms of arrays implies that

$$-\underline{A}^T \underline{f} = \underline{g} \qquad (8.17)$$

Recalling that the Newton-Raphson solution of eq. (8.8) is

$$p_{m+1} = p_m - \underline{G}_m^{-1} \underline{g}_m \tag{8.18}$$

$$p_{m+1} = p_m - (\underline{A}_m^T \cdot \underline{A}_m)^{-1} \underline{A}_m^T \underline{f}_m \tag{8.19}$$

which provides important information: the iteration of the parameter vector (p) is dependent on the previous function values (f) and the partial derivatives of the model equations.

The algorithm above, eq. (8.19), looks convenient as such for obtaining the parameters, but it suffers from a serious disadvantage: the convergence is guaranteed only in cases where the initial guess (p) is close enough to the actual solution. Therefore, the algorithm has been developed further. Levenberg has proposed a compromise algorithm between the Newton-Raphson and steepest descent algorithms:

$$\underline{p}_{m+1} = \underline{p}_m - (\lambda \underline{T} + \underline{A}_m^T \cdot \underline{A}_m)^{-1} \underline{A}_m^T f_m \tag{8.20}$$

$$\underline{T} = \begin{bmatrix} . & 0 & 0 \\ 0 & . & 0 \\ 0 & 0 & . \end{bmatrix}, \quad \lambda = scalar \tag{8.21}$$

If the scaling parameter, λ, is large, the steepest gradient method is obtained, but for the case where $\lambda = 0$, the Newton-Raphson method is obtained. Typically, the value of the scaling parameter (λ) is high at the beginning of the iteration but decreases as the minimum is approached. Marquardt (1963) developed a strategy for the selection of λ; therefore, the method is nowadays called Levenberg-Marquardt method.

8.2 Statistical and sensitivity analysis of parameters

An indicative and preliminary statistical analysis of the parameters obtained by regression analysis can be done by regarding two important quantities, namely the variances of the parameters and the correlation coefficients between them. These quantities are calculated from the objective function (Q) and the matrix of the parameter derivatives.

The matrix $\underline{A}$ is defined as

$$\underline{A} = \begin{bmatrix} \partial f_1/\partial p_1 & \partial f_1/\partial p_2 & \cdots & \partial f_1/\partial p_{np} \\ \partial f_2/\partial p_1 & \partial f_2/\partial p_2 & \cdots & \partial f_2/\partial p_{np} \\ \vdots & \vdots & \ddots & \vdots \\ \partial f_{np}/\partial p_1 & \partial f_2/\partial p_2 & \cdots & \partial f_{np}/\partial p_{np} \end{bmatrix} \tag{8.22}$$

and, for the sake of convenience, a matrix $\underline{L}$ is introduced:

$$\underline{L} = (\underline{A}^T \cdot \underline{A})^{-1} \tag{8.23}$$

The variances of parameters (p_k), σ_k^2, can be estimated from the objective function (Q) and from the diagonal elements of $\underline{L}$:

$$\sigma_k^2 = \frac{Q}{nm - np} L_{kk} \tag{8.24}$$

where nm and np denote the number of experiments and the number of parameters, respectively. The covariances of parameters p_k and p_l are defined by $\rho_{kl}\sigma_k\sigma_l$, where ρ_{kl} is the correlation coefficient between p_k and p_l.

The fundamental statistical relationship is

$$\rho_{kl}\sigma_k\sigma_l \approx \frac{Q}{nm - np} L_{kl} \tag{8.25}$$

On the other hand, the variances are given by

$$\sigma_k = \frac{\sqrt{Q}\sqrt{L_{kk}}}{\sqrt{nm - np}} \tag{8.26}$$

$$\sigma_l = \frac{\sqrt{Q}\sqrt{L_{ll}}}{\sqrt{nm - np}} \tag{8.27}$$

By combining the equations above, we get a simple expression for the correlation coefficient:

$$\sigma_k\sigma_l = \frac{Q}{nm - np} \sqrt{L_{kk}}\sqrt{L_{ll}} \tag{8.28}$$

The equation shows that the correlation coefficients are obtained from the elements of matrix $\underline{A}$,

$$\rho_{kl} = \frac{L_{kl}}{\sqrt{L_{kk}}\sqrt{L_{ll}}}, \qquad \rho_{kl} \in [0, 1] \tag{8.29}$$

Generally speaking, one of the goals in parameter estimation is to minimize variances as well as the correlation coefficients of the parameter, that is, σ_k should be as small as possible for each p_k. At the same time, ρ_{kl} should be small for each pair of parameters, $p_k - p_l$.

Small variances guarantee that the parameter is accurately estimated, and small correlation coefficients indicate that the parameters cannot be mutually compensated. The variances are very much dependent on the precision of the experiments, as they are directly proportional to the weighted sum of residual squares (Q), while the correlation coefficient depends heavily on the model structure as such. Special techniques to suppress the correlation between parameters exist and should always be used. They will be discussed in the following section.

Standard statistical analysis of parameters is based on linearization, and it does not always give a realistic picture of the accuracy of the parameters. For example, variance expresses the confidence interval of the parameter in a symmetric way, which sometimes can lead to misunderstandings: a small parameter, for example, a rate constant with a big variance, which means that even negative values of the parameter could be accepted. This, of course, is not true. A very simple and illustrative approach is to check the objective function (Q) as a function of each parameter value. All of the parameters except one are given the value corresponding with the minimum value of the objective function, Q, and the influence on the numerical value of the objective function, Q, is investigated as a function of the parameter under consideration. Examples of these kinds of plots are shown in Figure 8.2. Graphical considerations give additional information, for instance concerning symmetry: the parameter is often accurately defined in one direction but less accurately in the opposite one. This procedure is called sensitivity analysis.

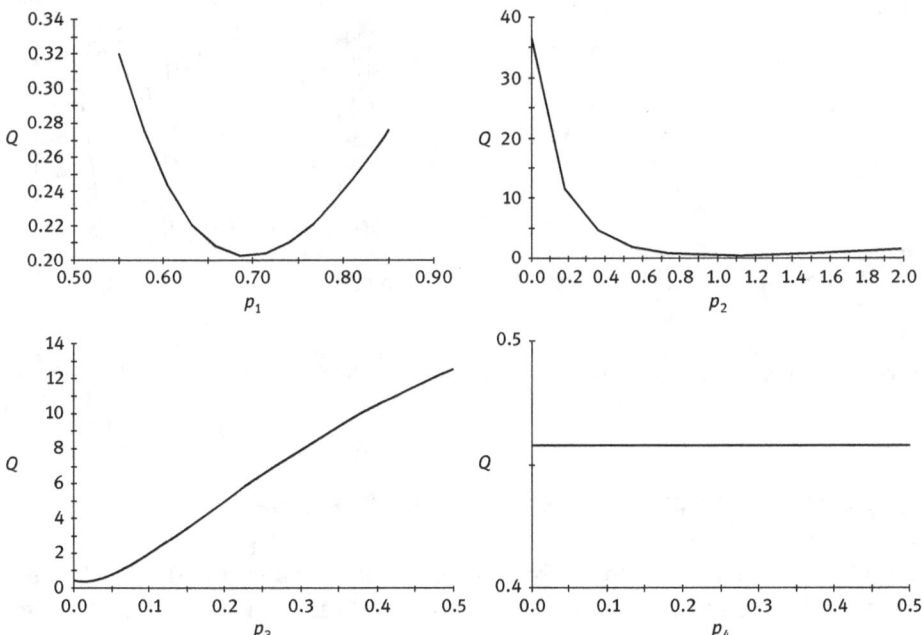

Figure 8.2: Parameter sensitivity analysis.

A standard way to describe the results of parameter estimation is a contour plot. The value of the objective function (Q) is investigated as a function of a pair of parameters, and the values where the objective function gets the same, specified value are depicted in a single plot. Examples of contour plots are provided in Figure 8.3. If correlation between the parameters is low, the contour plot consists of circles. In case of a strong mutual correlation between parameters, contour plots become elongated (Figure 8.3). These kinds of plots bear a clear message: the numerical value of one parameter can easily be compensated by another parameter, either by increasing or decreasing its value. In any case, we get an equally good fit to the experimental data. From statistical as well as physical-chemical viewpoint, this is not desirable: the exact values of the parameters remain uncertain, and the regression analysis has just provided a data fitting exercise. The model can be used for interpolation within the experimentally screened domain, but not for extrapolation outside it. Furthermore, the physical and chemical significances of the parameters remain questionable. The discussion shows that suppression of the correlation between parameters is a crucially important issue.

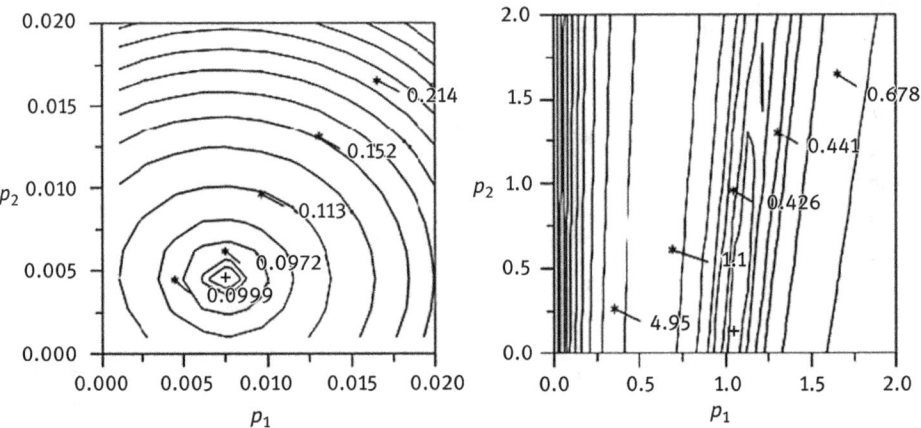

Figure 8.3: Contour plots of parameter pairs: (left) low correlation. (b) high correlation between the parameters.

8.3 Suppression of correlation between parameters

Correlation between estimated parameters (eq. 28) often is a serious drawback of the estimation: because of a strong correlation, numerical values of the individual parameters remain uncertain, even though the overall fit of the model to the experimental data might be rather good. Typical examples of strongly correlated parameters are the frequency factor and activation energy in the Arrhenius law for rate constants, as well as the kinetic and equilibrium parameters appearing in rate equations for catalytic processes. However, certain special techniques can be used to diminish the mutual correlation in regression analysis.

8.3.1 Correlation in rate expressions

Suppression of mutual correlation between parameters will be illustrated by a case study. Rates of catalytic reactions are very frequently measured in gradientless and differential reactors, and rate expressions of the Langmuir-Hinshelwood type are used in the interpretation of experimental data. The rate expression typically has the form

$$R = \frac{k_+ \left(\prod_{i=1}^{N^+} c_i^{\alpha_+ i} - \prod_{i=1}^{N^-} \frac{c_i^{\alpha_- i}}{K} \right)}{\left(1 + \sum K_j c_j^{\varphi_j} \right)^\delta} \tag{8.30}$$

where $N = N^+ + N^-$, that is, the number of components. N^- and N^+ denote the number of reactants and products, respectively. The values of the exponents depend on the reaction stoichiometry and reaction mechanism, and a more detailed discussion can be found in textbooks in the field of catalysis and reaction engineering. The equilibrium constant (K) is usually not a problem from the viewpoint of parameter estimation, since it can be determined separately from the equilibrium composition. The forward rate constant (k_+) and the adsorption parameters (K_i) are usually obtained from kinetic experiments by using non-linear regression analysis.

A serious drawback is the strong correlation between the rate constant (k_+) and the adsorption equilibrium parameters (K_i): if the value of the rate constant in the nominator of the rate expression is increased, its increase is compensated by the increase of the adsorption parameter, and we get an equally good fit to the experimental data. The correlation between parameters is illustrated with contour plots, as shown in Figure 8.3.

A general rule of thumb is that correlation between parameters can be diminished by transforming rational functions into sums. Consequently, by taking the reciprocal value and the root of the rate equation, we get the transformed rate expression for irreversible reactions $K \to \infty$

$$\frac{1}{R^\delta} = \frac{1}{k''} + \sum \frac{K_j}{k''} c_j^x \text{ where } k'' = k_+^{-1/\delta} (\ )^{-1/\delta} \tag{8.31}$$

for the case of irreversible kinetics.

Thus, a new set of parameters is defined by

$$\frac{1}{k''} = a_0, \quad \frac{K_j}{k''} = a_j, \quad j = 1 \ldots N \tag{8.32}$$

and the model is rewritten as

$$R = \frac{1}{\left(a_0 + \sum a_j x_j \right)^\delta} \tag{8.33}$$

where the new parameters are defined by

$$a_0 = \frac{1}{k_+^{1/\delta}}(\ )^\delta, \qquad a_j = \frac{K_j}{k_+^{1/\delta}}(\ )^\delta \qquad (8.34)$$

An additional advantage is that initial estimates of the parameters can be obtained from the rates by taking the reciprocal value of the equation,

$$\hat{y} = a_0 + \sum_{j=1}^{N} a_j x_j \qquad (8.35)$$

which de facto is a linear model with respect to the parameters $a_0 \ldots a_N$. The dependent variable is defined as

$$\hat{y} = \frac{1}{R^{1/\delta}} \qquad (8.36)$$

while the independent variables are obtained from the concentrations:

$$x_j = c_j^{x_j} \qquad (8.37)$$

Thus, we arrive at a model that is linear with respect to the parameters, and linear regression can be applied to it.

A simple illustration is considered: a bimolecular, catalytic reaction $A + B \rightarrow C$ obeys the rate expression

$$R = \frac{k c_A c_B}{(1 + K_A c_A + K_B c_B + K_C c_C)^2} \qquad (8.38)$$

The reciprocal square root leads to the following formulation

$$\frac{1}{\sqrt{R}} = \frac{1}{\sqrt{k}} \frac{1}{\sqrt{c_A c_B}} + \frac{K_A}{\sqrt{k}} \frac{\sqrt{c_A}}{\sqrt{c_B}} + \frac{K_B}{\sqrt{k}} \frac{\sqrt{c_B}}{\sqrt{c_A}} + \frac{K_C}{\sqrt{k}} \frac{\sqrt{c_C}}{\sqrt{c_A c_B}} \qquad (8.39)$$

The dependent variable is now

$$\hat{y} = \frac{1}{\sqrt{R}} \qquad (8.40)$$

while the independent variables are

$$x_1 = \frac{1}{\sqrt{c_A c_B}}, \quad x_2 = \frac{\sqrt{c_A}}{\sqrt{c_B}}, \quad x_3 = \frac{\sqrt{c_B}}{\sqrt{c_A}}, \quad x_4 = \frac{\sqrt{c_C}}{\sqrt{c_A c_B}} \qquad (8.41)$$

and the parameters to be estimated by regression analysis are:

$$a_1 = \frac{1}{\sqrt{k}}, \quad a_2 = \frac{K_A}{\sqrt{k}}, \quad a_3 = \frac{K_B}{\sqrt{k}}, \quad a_4 = \frac{K_C}{\sqrt{k}} \qquad (8.42)$$

The very simple linear model is thus obtained,

$$\hat{y} = \sum_j a_j x_j \tag{8.43}$$

that can be used in linear regression to get reasonable initial estimates of the final parameter estimation, which is carried out by non-linear regression, in accordance with the equation

$$\hat{y} = R = \frac{c_A c_B}{\left(a_1 + \sum_i a_i c_i\right)^2} \tag{8.44}$$

8.3.2 Correlation in temperature dependencies

The correlation between temperature-related parameters often severely affects the results of parameter estimation. A typical situation is visualized in Figure 8.4. The process model follows the format,

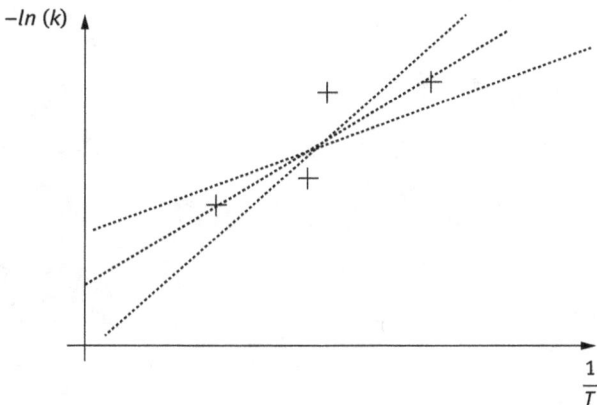

Figure 8.4: Correlation between parameters.

$$y = p_0 + p_1 x \tag{8.45}$$

This is in principle a very simple model, but experiments are in practise limited to a narrow interval of x and thus the value of p_0 is easily compensated by p_1 and vice versa. This is the case when the frequency factor (A) and the activation energy (E_a) of the rate constant are determined from the famous law of Arrhenius:

$$k = A \cdot e^{\frac{-E_a}{RT}} \tag{8.46}$$

The values of the rate constant, k, are obtained within a certain temperature interval, and some experimental scattering typically appears in the k values. The challenge is illustrated by Figure 8.4. In a logarithmic plot, $ln(k)$ versus $1/T$ gives

$$ln(k) = ln(A) - \frac{E_a}{RT} \tag{8.47}$$

The intercept ($ln(A)$) can be compensated by the slope ($-E_a/R$), and both parameters remain uncertain. In general, it is possible to obtain a reasonably good fit to the k values, but individual parameter values (A and E_a) can remain highly uncertain. The same effect appears in the estimation of reaction enthalpy from the equilibrium constants measured at different temperatures, that is, in the application of the van't Hoff law: k and E_a in eq. (8.47) correspond to K and ΔH_r in the law of van't Hoff. One way to improve the parameter estimation procedure is to orthogonalize (centralize) the experiments according to the procedure described in Figure 8.5.

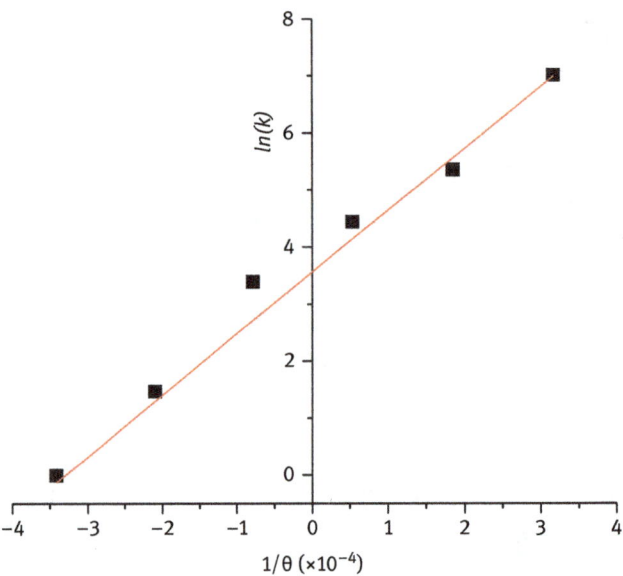

Figure 8.5: The rate constant according to the Arrhenius law.

The abscissa axis is shifted to the average value of the experiments (x), $x' = x - \bar{x}$. After applying this to the Arrhenius law, we obtain

$$\frac{1}{\theta} = \frac{1}{T} - \frac{1}{\bar{T}} \tag{8.48}$$

The problem is reformulated as

$$k = \bar{k} \cdot e^{-\frac{E_a}{R}\frac{1}{\theta}} \qquad (8.49)$$

where $\bar{k} = A \cdot exp(-\frac{E_a}{RT})$, that is, the rate constant at the average temperature. The advantages of this restructuring are evident: the experiments are centralized, suppressing correlation between the parameters. Furthermore, a very abstract parameter, the frequency factor, is replaced by the rate constant at the average temperature ($\bar{k}$). It is much easier to make an initial guess of this parameter, the rate constant, rather than of the frequency factor in non-linear regression.

However, the lesson learnt from this particular example, Arrhenius law, is general. Whenever the mathematical model has this kind of a structure, correlation between parameters is suppressed by orthogonalization of the experimental domain, that is, by taking averages and defining a new coordinate,

$$x' = x - \bar{x} \qquad (8.50)$$

inserting it in the original model yields

$$\hat{y} = a_0 + a_1 x \qquad (8.51)$$

$$\hat{y} = (a_0 + a_1 \bar{x}) + a_1 x' \qquad (8.52)$$

A new parameter,

$$a_0' = a_0 + a_1 \bar{x} \qquad (8.53)$$

is defined and the model becomes

$$\hat{\bar{y}} = \hat{y} - \bar{y} = \hat{y} - (a_0 + a_1 \bar{x}) \qquad (8.54)$$

$$\hat{\bar{y}} = \hat{y} - \bar{y} = a_1 x' \qquad (8.55)$$

The estimation procedure can be summarized as follows:
1) The values of x and y are calculated as averages from the experimental data.
2) The values of x_i' are obtained as in eq. (8.50).
3) Linear regression is performed, giving $a_0' = a_0 + a_1 \bar{x}$ and a_1 as the initial estimates.
4) The real parameter a_0 is obtained.

8.4 Systematic deviations and normalization of experimental data

Not all deviations between the model and the data are of a stochastic nature. Analytical techniques, particularly automation of offline analysis, such as gas and liquid chromatography, and development of online analytical techniques (UV, FTIR,

FIA, SIA) reduce the random scattering in the data to a minimum, and beautiful experimental curves can be plotted.

Still, a number of deviations appear between experimental and predicted data. The main reason originates from systematic deviations, which are easily recognized in a graphical consideration of data sets (Figure 8.6). Systematic deviations are caused by several reasons; the most common ones being inadequate stoichiometry, an inappropriate kinetic model and poor calibration of analytical equipment. That the kinetic model is 'wrong' is easily recognized in graphical plots; for instance, an erroneous reaction order can be seen by comparing the curvature of the experimental data and predicted points as visualized in Figure 8.6. The next steps to take are obvious: an improved kinetic model is applied to the data, and a new comparison is made. The iteration is continued until an adequate fit is obtained.

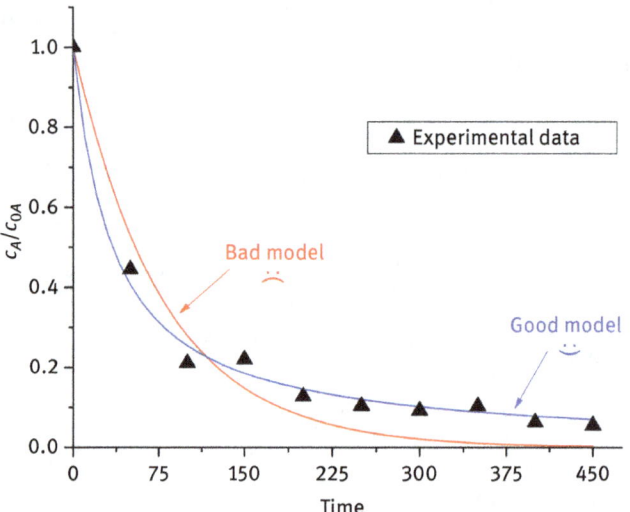

Figure 8.6: Systematic deviation due to a poorly chosen kinetic model: good model: second-order kinetics, "bad" model: 1st order kinetics.

Inadequate stoichiometry and poor calibration of the analytical device are interconnected problems. The kinetic model itself follows stoichiometric rules, but an inadequate calibration of the analytical instrument causes systematic deviations. This can be illustrated with a simple example. Let us assume that a bimolecular reaction, $A + B \rightarrow P$, is carried out in a homogeneous liquid-phase batch reactor. The density of the reaction mixture is assumed to be constant. The reaction is started with A and B, and no P is present in the initial mixture. The concentrations are related by

$$c_P = c_{0A} - c_A = c_{0B} - c_B \tag{8.56}$$

that is, the produced product, P, equals the consumed reactant. If the concentration of component B has a calibration error, instead of the correct concentration c_B, we get an erroneous one,

$$c_B' = \alpha c_B \tag{8.57}$$

which does not fulfil the stoichiometric relation. If the error is large for a single component, it is easy to recognize, but the situation can be much worse: calibration errors are present in several components and all of their effects are spread during non-linear regression in the estimation of model parameters. This is reflected by the fact that the total mass balance is not fulfilled by the experimental data. A way to check the analytical data is to use some form of total balances, for example, atom balances or total molar amounts or concentrations. For example, for our model reaction, $A + B \to P$, we have the relations

$$c_A + c_P = c_{0A} = \text{constant} \tag{8.58}$$

$$c_B + c_P = c_{0B} = \text{constant} \tag{8.59}$$

$$c_A + c_B + 2c_P = c_{0A} + c_{0B} = \text{constant (again } c_{0P} = 0) \tag{8.60}$$

The sum of concentrations of A and P should remain constant in space and time; if not, something is wrong (Figure 8.7). Either the reaction does not follow this stoichiometry or – more probably – the analysis is not calibrated very exactly. The drawback of considering the total balance is that a lot of information is lost, since many concentrations are merged together.

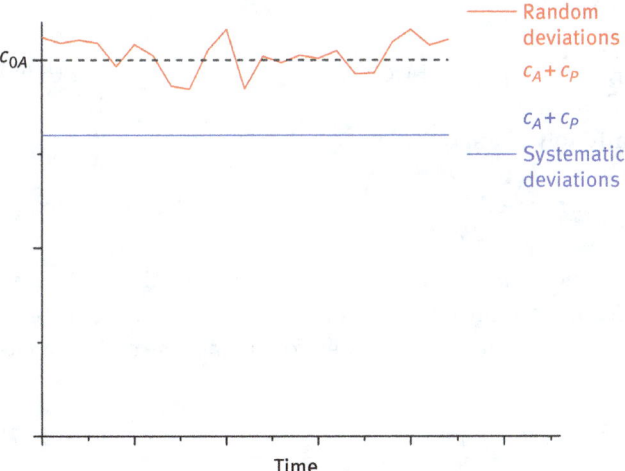

Figure 8.7: Checking kinetic data against total balance.

A more advanced way of considering potential calibration errors is to look at the components individually. An algorithm which can be used for checking kinetic data and stoichiometric normalization of data is presented here.

A single reaction is considered. The extent of the reaction is defined by

$$\xi = \frac{y_i - y_{0i}}{\nu_i} \quad (8.61)$$

where ξ is the extent of reaction, y_i denotes the molar quantity after some reaction or residence time in the reactor, and y_{0i} is the initial or inlet quantity. For a continuous reactor, y_i is the molar flow, for batch reactors, it is the molar amount of substance or concentration. For systems with a constant density, the molar quantities can simply be replaced by concentrations. Thus, any model predicts the stoichiometric relation

$$\hat{y}_i = y_{0i} + \nu_i \xi \quad (8.62)$$

On the other hand, experimentally recorded concentrations exist for components in the system. The task is now to find an optimal extent of reaction that would minimize the difference for the entire data set. The objective function is defined as

$$Q = \sum_i w_i (\hat{y}_i - y_i)^2 \quad (8.63)$$

where w_i is the weight factor for component i. The stoichiometric relationship, eq. (8.62), is inserted in eq. (8.63) giving

$$Q = \sum_i w_i (y_{0i} + \nu_i \xi - y_i)^2 \quad (8.64)$$

An optimal value for the extent of the reaction, ξ, is found by differentiation of eq. (8.64) with respect to ξ,

$$\frac{dQ}{d\xi} = \sum_i 2 w_i \nu_i (y_{0i} + \nu_i \xi - y_i) = 0 \quad (8.65)$$

in which ξ can be conveniently solved analytically:

$$\xi = \frac{\sum w_i \nu_i (y_i - y_{0i})}{\sum w_i \nu_i^2} \quad (8.66)$$

Thus, stoichiometrically consistent values of molar quantities are obtained by back-substituting the expression for the optimized extent of reaction, ξ, in the basic relation, eq. (8.62). For an arbitrary component (k), we therefore obtain the stoichiometrically corrected value:

$$y_k = y_{0k} + \nu_k \xi \quad (8.67)$$

where ξ is calculated from eq. (8.66). The consistency of the data can now be checked by plotting the initial value (c_i) and the corrected value (c_k) in the same graph. Different weight factors are used if it is a priori known that some of the components are better calibrated than others. Furthermore, the weight factors can be used for testing purposes; it is possible to try it out and see what happens if we rely on some part of the data very heavily, and how much the other concentrations change by this stoichiometric normalization.

The procedure presented above for a single-reaction case is easily extended to several simultaneous reactions. For simultaneous reactions, the stoichiometric relationship is written as

$$\underline{\hat{y}} = \underline{y_{0k}} + v_k \xi \tag{8.68}$$

The objective function is defined analogously to eq. (8.63),

$$Q = \sum_i w_i (\hat{y}_i - y_i)^2 \tag{8.69}$$

and the derivatives with respect to each extent of reaction are set to zero, that is,

$$\frac{dQ}{d\xi_j} = \sum_i^N 2 \left(y_i - y_{0i} - \sum_{j=1}^S \xi_j v_{ij} \right) (-v_i) w_i = 0 \tag{8.70}$$

Finally, it can be shown that the corrected quantities are obtained in the algorithm

$$\underline{\xi} = \underline{A}^{-1} \underline{B}, \quad \underline{y} = \underline{y}_0 + v \underline{A}^{-1} \underline{B} \tag{8.71}$$

where v is the stoichiometric matrix and the elements of the array, A and B, are calculated from

$$\underline{A} = \begin{bmatrix} v_1^2 & v_1 v_2 & \cdots & v_1 v_S \\ v_2 v_1 & v_2^2 & \cdots & v_2 v_S \\ \vdots & \vdots & \ddots & \vdots \\ v_S v_1 & v_S v_2 & \cdots & v_S^2 \end{bmatrix} \tag{8.72}$$

$$\underline{B} = \begin{bmatrix} \sum w_i (y_i - y_{0i}) v_{i1} \\ \vdots \\ \sum w_i (y_i - y_{0i}) v_{iS} \end{bmatrix} \tag{8.73}$$

The procedure can be easily implemented by a computer.

An example of the application of the algorithm is given in Figure 8.8. Hydrogenation of D-xylose to xylitol was carried out in a batchwise operating slurry reactor. The analytical data was obtained by means of high-performance liquid chromatography (HPLC). The original data did not obey the stoichiometry

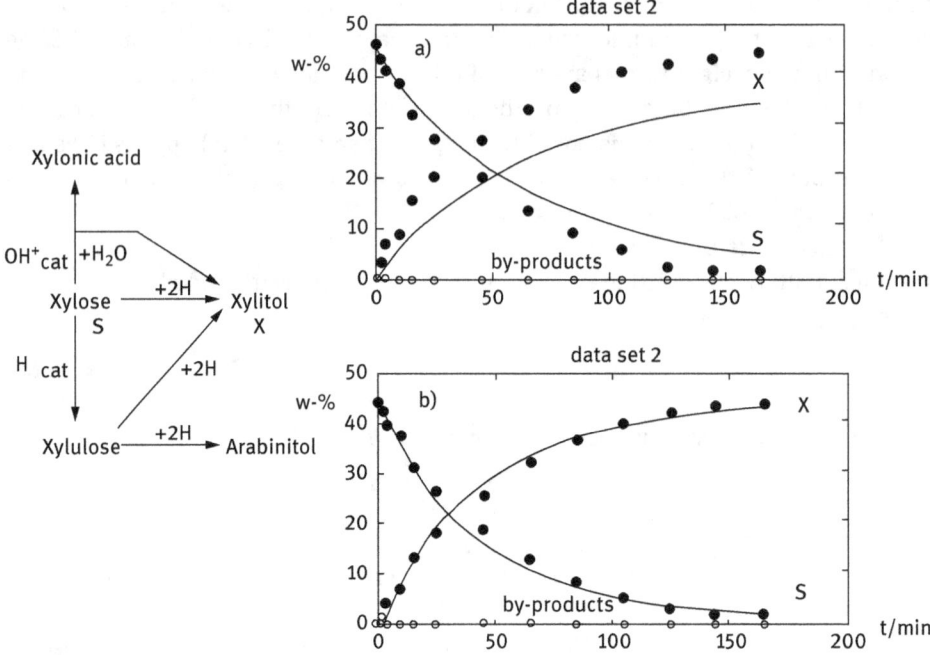

Figure 8.8: Results of data normalization: fit of original (a) and normalized (b) data. Case study: hydrogenation of D-xylose to xylitol.

exactly due to calibration problems typically associated with this analytical technique. This implies that systematic deviations remain in the estimation of kinetic parameters, and the fit of the model parameters is not very good (Figure 8.8). At the next stage, the original data were exposed to the normalization algorithm, eq. (8.73), and the parameter estimation was repeated with stoichiometrically consistent data. The systematic deviations disappeared, and a much better fit was obtained (Figure 8.8). It should be emphasized that the final goal should be the improvement of the primary data to remove the inconsistencies of calibration that distort the mass balance. The presented algorithm is just a diagnostic tool that can be used to localize deviations and investigate their effect on parameter estimation.

8.5 Direct integral method

A special method which can essentially simplify the regression analysis procedure is presented here, namely the direct integral method (Himmelblau et al. (1967)). For a given chemical reactor, we can often present a kinetic model as a set of ordinary differential equations of the form:

8.5 Direct integral method

$$\frac{dc_i}{d\tau} = \rho \sum_j v_{ij} \cdot R_j \tag{8.74}$$

where v_{ij} is the stoichiometric coefficient, R_j the reaction rate, and ρ is a proportionality factor (e.g. catalyst bulk density for catalytic processes; for homogeneous reactors $\rho = 1$). Finally, τ is the reaction time (batch reactor) or the space time (continues reactor). Furthermore, the kinetic model is often presented as

$$\frac{dc_i}{d\tau} = \rho \sum_j v_{ij} \cdot \left(k_{+j}\Pi_j^+ - k_{-j}\Pi_j^-\right) \tag{8.75}$$

where $\Pi_j^{+/-} = \Pi_k c_k^{\alpha_{kj}^{+/-}}$ and α_{kj} are the reaction exponents with respect to different compounds. These exponents α_{kj} can correspond with stoichiometric coefficients (elementary reactions) or with empirical figures deduced from experimental data. Formally, a direct numerical integration of the normalized experimental data gives the solution of the ODEs in the form:

$$\hat{c}_{t,i} = c_{0,i} + \rho \int_0^{\tau_t} \sum_j v_{ij} \cdot \left(k_{+j}\Pi_j^+ - k_{-j}\Pi_j^-\right) d\tau \tag{8.76}$$

where τ_t denotes the reaction or space time for which the molar quantity $\hat{c}_{t,i}$ was obtained. This solution is then inserted in the objective function to obtain the following expression:

$$Q = \sum_t \sum_i \left(c_{0,i} + \rho \int_0^{\tau_t} \sum_j v_{ij} \cdot \left(k_{+j}\Pi_j^+ - k_{-j}\Pi_j^-\right) d\tau - c_{t,i}\right)^2 \omega_{t,i} \tag{8.77}$$

which is rearranged as

$$Q = \sum_t \sum_i \left(c_{0,i} + \rho \sum_j v_{ij} \cdot k_{+j} \int_0^{\tau_t} \Pi_j^+ d\tau - \rho \sum_j v_{ij} \cdot k_{-j} \int_0^{\tau_t} \Pi_j^- d\tau - c_{t,i}\right)^2 \omega_{t,i} \tag{8.78}$$

The expression is further differentiated with respect to kinetic parameters (k), and the derivatives are set to zero to obtain the optimal values of the kinetic parameters:

$$\frac{dQ}{dk_l^+} = 2\sum_t \sum_i \left(c_{0,i} + \rho \sum_j v_{ij} \cdot k_{+j} \int_0^{\tau_t} \Pi_j^+ d\tau - \rho \sum_j v_{ij} \cdot k_{-j} \int_0^{\tau_t} \Pi_j^- d\tau - c_{t,i}\right)$$
$$\omega_{t,i} \cdot \rho \cdot v_{il} \int_0^{\tau_t} \Pi_l^+ d\tau \tag{8.79}$$

$$\frac{dQ}{dk_l^-} = -2\sum_t\sum_i \left(c_{0,i} + \rho\sum_j v_{ij}\cdot k_{+j}\int_0^{\tau_t}\Pi_j^+ d\tau - \rho\sum_j v_{ij}\cdot k_{-j}\int_0^{\tau_t}\Pi_j^- d\tau - c_{t,i}\right) \quad (8.80)$$

$$\omega_{t,i}\cdot\rho\cdot v_{il}\int_0^{\tau_t}\Pi_l^- d\tau$$

After numerous algebraic steps, a rearrangement of eqs. (8.79) and (8.80) gives the following expression. The subindex l varies from 1 to NR, where NR represents the number of chemical reactions in the studied system.

$$\sum_t\left[\sum_j\left(\sum_i v_{ij}\int_0^{\tau_t}\Pi_j^+ d\tau \cdot v_{il}\int_0^{\tau_t}\Pi_l^+ d\tau \cdot \omega_{t,i}\cdot k_{+j}\right)\right.$$
$$\left.-\sum_j\left(\sum_i v_{ij}\int_0^{\tau_t}\Pi_j^- d\tau \cdot v_{il}\int_0^{\tau_t}\Pi_l^+ d\tau\, \omega_{t,i}\cdot k_{-j}\right)\right] \quad (8.81)$$
$$= \sum_t\left(\sum_i (c_{0i} - c_i)\cdot\omega_{t,i}\cdot v_{ij}\cdot\int_0^{\tau_t}\Pi_l^+ d\tau\right)\Big/\rho$$

$$\sum_t\left[\sum_j\left(\sum_i v_{ij}\int_0^{\tau_t}\Pi_j^+ d\tau \cdot v_{il}\int_0^{\tau_t}\Pi_l^- d\tau \cdot \omega_{t,i}\cdot k_{+j}\right)\right.$$
$$\left.-\sum_j\left(\sum_i v_{ij}\int_0^{\tau_t}\Pi_j^- d\tau \cdot v_{il}\int_0^{\tau_t}\Pi_l^- d\tau\, \omega_{t,i}\cdot k_{-j}\right)\right] \quad (8.82)$$
$$= \sum_t\left(\sum_i (c_{0i} - c_i)\cdot\omega_{t,i}\cdot v_{ij}\cdot\int_0^{\tau_t}\Pi_l^+ d\tau\right)\Big/\rho$$

The next step would be to set the values of j to integers in the range $(1, NR)$, where NR represents the number of reactions in the system. Setting the value of $j = 1$, one obtains:

$$\sum_t\left[\left(\sum_i v_{il}\cdot\int_0^{\tau_t}\Pi_l^+ d\tau \cdot v_{il}\cdot\int_0^{\tau_t}\Pi_l^+ d\tau \cdot \omega_{t,i}\cdot k_{+j}\right)\right.$$
$$\left.-\left(\sum_i v_{il}\cdot\int_0^{t}\Pi_j^- d\tau \cdot v_{il}\cdot\int_0^{\tau_t}\Pi_l^+ d\tau\cdot \omega_{t,i}\cdot k_{-j}\right)\right] \quad (8.83)$$
$$= \sum_t\left(\sum_i (c_{0i} - c_i)\cdot\omega_{t,i}\cdot v_{ij}\cdot\int_0^{\tau_t}\Pi_l^+ d\tau\right)\Big/\rho$$

$$\sum_{t}\left[\left(\sum_{i} v_{il} \cdot \int_{0}^{\tau_t} \Pi_l^+ \, d\tau \cdot v_{il} \cdot \int_{0}^{\tau_t} \Pi_l^+ \, d\tau \cdot \omega_{t,i} \cdot k_{+j}\right)\right.$$
$$\left. - \left(\sum_{i} v_{il} \cdot \int_{0}^{\tau_t} \Pi_j^- \, d\tau \cdot v_{il} \cdot \int_{0}^{\tau_t} \Pi_l^+ \, d\tau \cdot \omega_{t,i} \cdot k_{-j}\right)\right] \tag{8.84}$$
$$= \sum_{t}\left(\sum_{i}(c_{0i} - c_i) \cdot \omega_{t,i} \cdot v_{ij} \cdot \int_{0}^{\tau_t} \Pi_l^+ \, d\tau\right)\Big/\rho$$

Equation (8.83) represents a general expression of the differentiation with respect to the forward reaction constants for each of the reactions involved in the system, while Eq. (8.84) represents differentiation with respect to the backward ones. The index l varies from 1 to NR. The value of Π_l will depend on the variable for which the differentiation was carried out (k_l). The procedure then continues, as we fix j for the successive integers, that is, $j = 2, 3, \ldots, NR$.

As the above equations indicate, it is possible to obtain a general linear expression with respect to the rate parameters (k_{+j}, k_{-j}) as follows,

$$a_{1,1}k_1 - a_{-1,1}k_{-1} + a_{2,1}k_2 - a_{-2,1}k_{-2} + \cdots + a_{NR,1}k_{NR} - a_{-NR,1}k_{-NR} = b_1$$
$$a_{1,-1}k_1 - a_{-1,-1}k_{-1} + a_{2,-1}k_2 - a_{-2,-1}k_{-2} + \cdots + a_{NR,-1}k_{NR} - a_{NR,-1}k_{-NR} = b_{-1}$$
$$a_{1,2}k_1 - a_{-1,2}k_{-1} + a_{2,2}k_2 - a_{-2,2}k_{-2} + \cdots + a_{NR,2}k_{NR} - a_{-NR,2}k_{-NR} = b_2$$
$$a_{1,-2}k_1 - a_{-1,-2}k_{-1} + a_{2,-2}k_2 - a_{-2,-2}k_{-2} + \cdots + a_{NR,2}k_{NR} - a_{NR,-2}k_{-NR} = b_1$$
$$\vdots$$
$$a_{1,NR}k_1 - a_{-1,NR}k_{-1} + a_{2,NR}k_2 - a_{-2,NR}k_{-2} + \cdots + a_{NR,NR}k_{NR} - a_{-NR,NR}k_{-NR} = b_{NR}$$
$$a_{1,-NR}k_1 - a_{-1,-NR}k_{-1} + a_{2,-NR}k_2 - a_{-2,-NR}k_{-2} + \cdots + a_{NR,-NR}k_{NR} - a_{-NR,-NR}k_{-NR} = b_{NR} \tag{8.85}$$

The values of a_{jl} and b_j are obtained from

$$a_{j,l} = \sum_{t}\left(\sum_{i} v_{ij} \cdot \int_{0}^{\tau_t} \Pi_j^+ \, d\tau \cdot v_{il} \cdot \int_{0}^{\tau_t} \Pi_l^+ \, d\tau \cdot \omega_{t,i}\right) \tag{8.86}$$

$$a_{-j,l} = \sum_{t}\left(\sum_{i} v_{ij} \cdot \int_{0}^{\tau_t} \Pi_j^- \, d\tau \cdot v_{il} \cdot \int_{0}^{\tau_t} \Pi_l^+ \, d\tau \cdot \omega_{t,i}\right) \tag{8.87}$$

$$a_{j,-l} = \sum_{t}\left(\sum_{i} v_{ij} \cdot \int_{0}^{\tau_t} \Pi_j^+ \, d\tau \cdot v_{il} \cdot \int_{0}^{\tau_t} \Pi_l^- \, d\tau \cdot \omega_{t,i}\right) \tag{8.88}$$

$$a_{-j,-l} = \sum_t \left(\sum_i v_{ij} \cdot \int_0^{\tau_t} \Pi_j^- \, d\tau \cdot v_{il} \cdot \int_0^{\tau_t} \Pi_l^- \, d\tau \cdot \omega_{t,i} \right) \tag{8.89}$$

$$b_l = \sum_t \left(\sum_i (c_{0i} - c_i) \cdot \omega_{t,i} \cdot v_{il} \cdot \int_0^{\tau_t} \Pi_l^+ \, d\tau \right) \Big/ \rho \tag{8.90}$$

The value of the integrals can be determined by natural splines approximation functions of the proposed kinetic model. The result of all this manipulation is a linear system of $2 \cdot NR$ algebraic equations with $2 \cdot NR$ unknowns that can be solved by any standard method, for example, Gauss-Jordan elimination. In case of irreversible chemical reactions only, the system becomes one with NR equations with NR unknowns. It can be written in matrix form as follows:

$$A = \begin{pmatrix} a_{1,1} & a_{-1,1} & a_{2,1} & a_{-2,1} & \cdots & a_{NR,1} & a_{-NR,1} \\ a_{1,-1} & a_{-1,-1} & a_{2,-1} & a_{-2,-1} & \cdots & a_{NR,-1} & a_{-NR,-1} \\ a_{1,2} & a_{-1,2} & a_{2,2} & a_{-2,2} & \cdots & a_{NR,2} & a_{-NR,2} \\ a_{1,-2} & a_{-1,-2} & a_{2,-2} & a_{-2,-2} & \cdots & a_{NR,-2} & a_{-NR,-2} \\ \vdots & \vdots & \vdots & \vdots & \ddots & \vdots & \vdots \\ a_{1,NR} & a_{-1,NR} & a_{2,NR} & a_{-2,NR} & \cdots & a_{NR,NR} & a_{-NR,NR} \\ a_{1,-NR} & a_{-1,-NR} & a_{2,-NR} & a_{-2,-NR} & \cdots & a_{NR,-NR} & a_{-NR,-NR} \end{pmatrix} \quad b = \begin{pmatrix} b_1 \\ b_{-1} \\ b_2 \\ b_{-2} \\ \vdots \\ b_{NR} \\ b_{-NR} \end{pmatrix} \tag{8.91}$$

and the values of the kinetic constants are then calculated from

$$\underline{k} = \underline{A}^{-1} \cdot \underline{b} \tag{8.92}$$

In this sense, starting from a proposed kinetic model and the stoichiometry of complex reactions, one can reduce a complex parameter estimation problem to a system of linear equations. The method is useful particularly for large reaction systems in which it is a challenge to obtain reasonable initial estimates for the kinetic parameters. A direct integral method should be used first to get very good parameter estimates, and these can be further improved by applying software for non-linear regression to the problem and numerical solution of the differential equations. We should, however, keep in mind the main limitation of the approach presented above: a requirement for using it is that all components are analysed chemically and that the power-law rate eq. (8.75) can be applied to the system.

8.6 Parameter estimation from non-isothermal data

Chemists traditionally think that kinetic data should be obtained under isothermal conditions. The reactants are preheated separately and brought together for the desired reaction to proceed. However, as solid catalysts are involved, the practical procedure becomes more complicated. Which component should the catalyst be combined with during preheating and how to pre-treat the catalyst? Does the pretreatment procedure contribute to catalyst deactivation? – These are typical questions raised when handling solid catalysts together with gases and liquids.

In practice, many experiments are carried out in such a way that all of the reactants are simply brought together with the catalyst, the mixture is heated up to the reaction temperature, and while being heated, the mixture undergoes at least some reaction under the rising temperature gradient to the present temperature. Nowadays this is not a technical problem, since the temperature profile can be followed on-line and stored on a computer as demonstrated in Figure 8.9. The temperature dependence of the kinetic model is incorporated in the temperature dependencies of the rate and equilibrium constants, for example, according to the laws of van't Hoff and Arrhenius. All data sets are merged together, and the parameters are determined by non-linear regression. Preferably, techniques to suppress mutual correlation between the parameters should be used. As an example, determination of parameters from hydrogenation experiments is shown in Figure 8.9 (Russo et al. (2015)). As can be seen in this Figure, a very good model agreement with the original data was achieved and the parameters fulfilled all significant standard tests.

A further challenge appears if the reactor equipment itself is non-isothermal. A typical example of this kind is a catalytic fixed bed. If the reaction comprises considerable heat effects, for instance highly exothermal reactions, it is difficult to maintain completely isothermal conditions in a tubular fixed bed. Instead, a hot spot appears, as illustrated by the high-temperature water-gas shift reaction in Figure 8.10.

One way to surmount this dilemma is to place thermocouples along the reactor tube and to screen the exact temperature field. Later on, the temperature profiles can be incorporated in the reactor model. Principally, two possibilities exist: to describe the reactor with a complete non-isothermal model including the energy balance, or to utilize the experimental temperature profiles more or less directly. The first approach is theoretically attractive, but it has the drawback that the heat transfer parameters of the reactor tube need to be determined. This introduces an additional uncertainty factor in the model, and the bottom line is that it is not relevant if the kinetic and thermodynamic parameters are of primary interest. Thus, it is better to interpolate in the experimental temperature field, for example by fitting an empirical model to each temperature profile, $T = f(\text{reactor length})$. A polynomial model usually is sufficient. The laws of van't Hoff and Arrhenius are incorporated in the thermodynamic and kinetic parameters, all data sets are merged together with the empirical temperature profiles, and non-linear regression is allowed to progress.

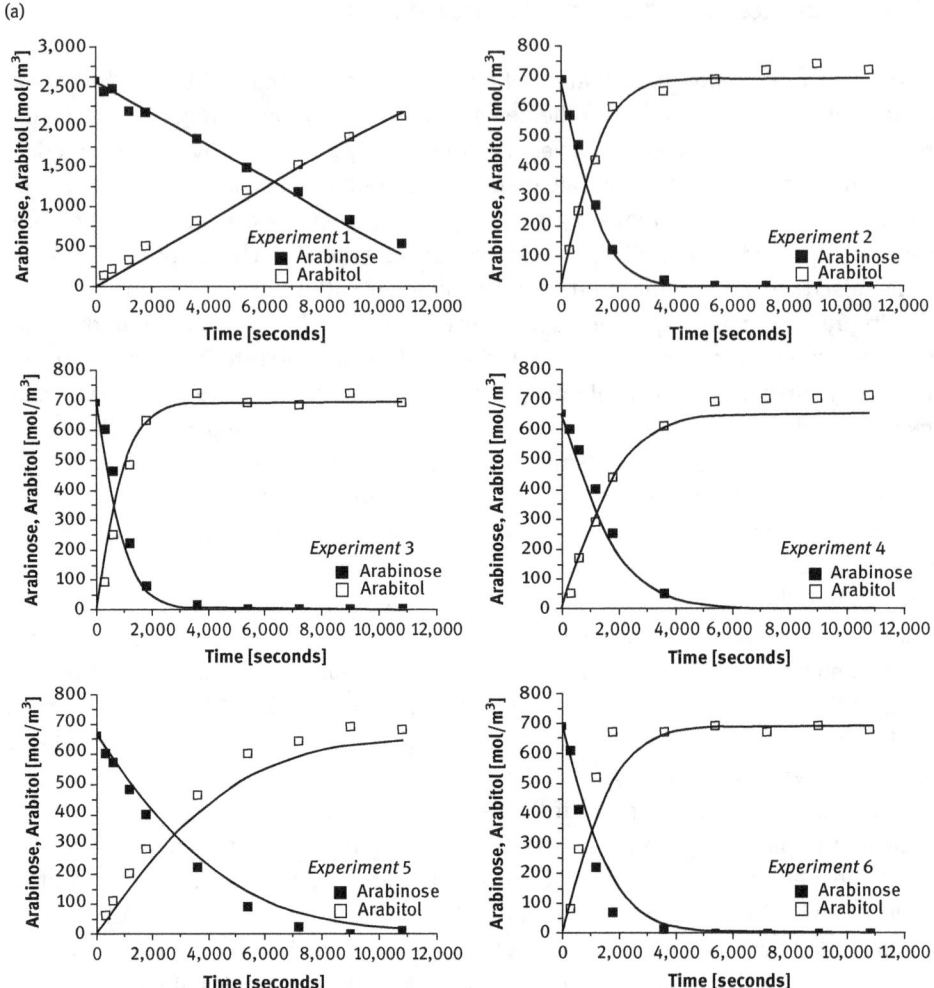

Figure 8.9: Non-isothermal experiments in a trickle bed reactor with internal diffusion and heat conduction. Case study: sugar hydrogenation (Russo et al. (2015)).

As an example, fitting a kinetic model to a water-gas shift reactor in a non-isothermal fixed bed reactor is shown in Figure 8.10 (Keiski et al. (1992)). The example demonstrates clearly that non-isothermal data indeed is feasible for kinetic analysis. The crucially important issue is to obtain primary data with a large enough concentration and temperature domain, and to record the temperature profiles frequently and precisely. As these requirements are fulfilled, the parameter estimation exercise can be safely carried out under non-isothermal conditions.

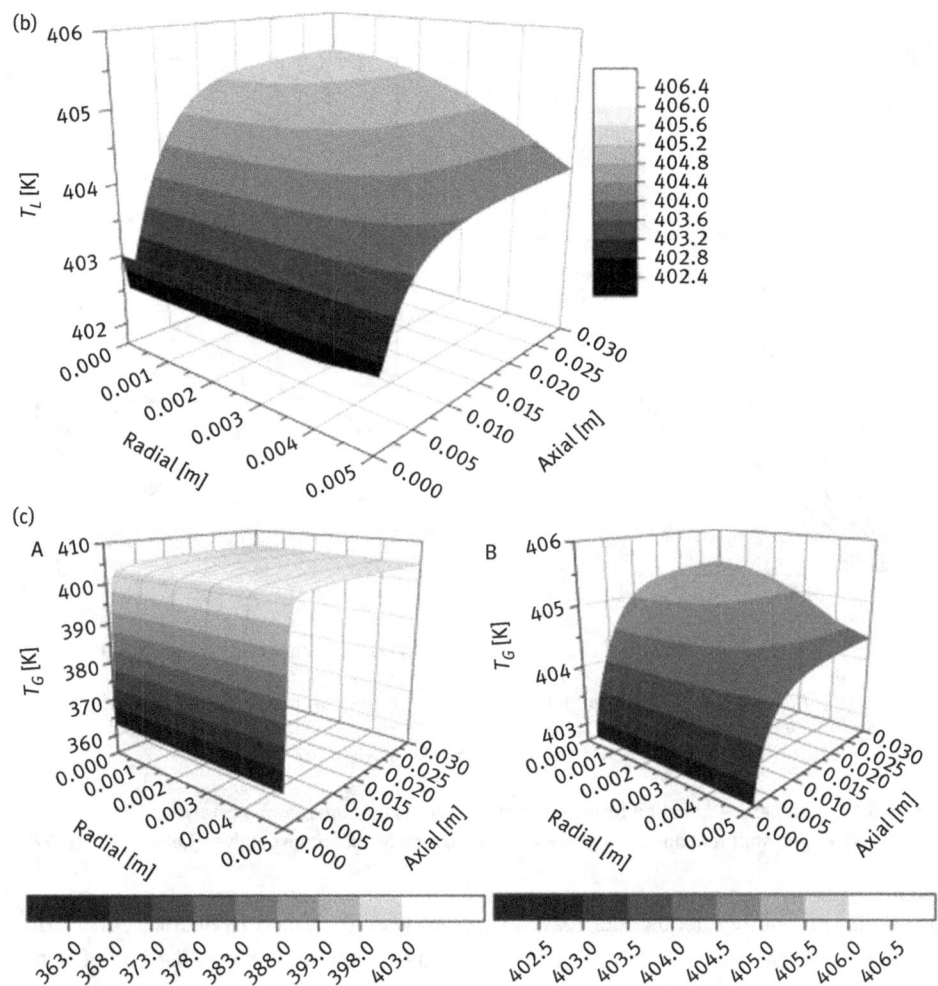

Figure 8.9 (continued)

8.7 Estimation of parameters from semi-batch experiments

Composite reactions appear in numerous chemical systems and in many sectors of chemical industry, from atmospheric chemistry to processes designed for fine and specialty chemicals, in oil refining and petrochemistry, in treating minerals and in biomass valorization. The general problem is that a simultaneous and reliable estimation of kinetic parameters is very demanding for this kind of multi-reaction systems, because the kinetic parameters are often strongly coupled and a multidimensional search in non-linear regression analysis can easily fail. Semi-batch technology can in some cases be helpful in improving the parameter estimation results. The general

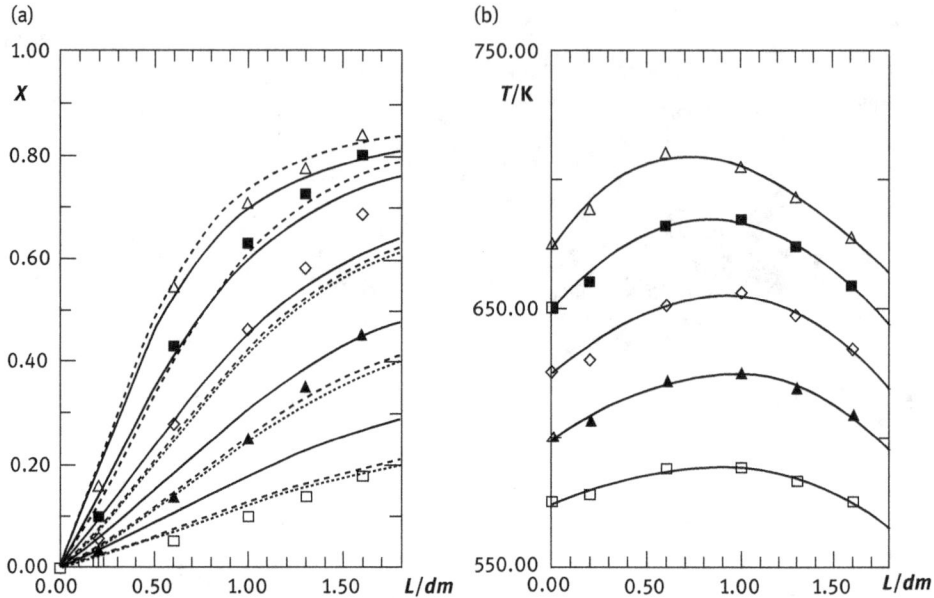

The model predictions compared with (a) independent conversion and (b) temperature data: □, T_0 = 575 K; ▲, T_0 = 600 K; ◇, T_0 = 625 K; ■, T_0 = 650 K; △, T_0 = 675 K; ——, two-parameter model; ·····, three-parameter model; – – –, four-parameter model. The inlet conditions were $\dot{n}_{0,co}$ = 1.1 mol h^{-1} and S = 1997 h^{-1}; the inlet mole ratios were H_2O/CO = 6.1, CO_2/CO = 0.67 and H_2/CO = 4.0.

Figure 8.10: Parameter estimation from non-isothermal data obtained from a catalytic fixed bed reactor. Water-gas shift reaction $CO + H_2O = CO_2 + H_2$ over a ferrochrome catalyst (Keiski et al. (1992)).

principles of homogeneous semi-batch reactors was discussed in Section 3.1. In this section, two special cases are considered in more detail: the application of semi-batch technology on heterogeneously and homogeneously catalysed liquid-phase systems, where several reactions proceed simultaneously.

8.7.1 Composite reactions in the presence of a heterogeneous catalyst

Estimation of kinetic parameters for complex consecutive and consecutive-competitive reactions of the type

$$A \rightarrow R \rightarrow S \rightarrow T \rightarrow \ldots$$
$$+B \quad +B \quad +B$$

is challenging not only because many kinetic parameters (rate and equilibrium constants) are involved, but the rates of the reactions can be very different. For instance, in the hydrogenation and oxidation of polyfunctional molecules, the first reaction steps are rapid, but the rates of the subsequent steps are much more slow, $R_1 \gg R_2, R_3, R_4 \ldots$ A typical set of kinetic curves are displayed in Figure 8.11, which describes the kinetics of citral hydrogenation with molecular hydrogen in the presence of a solid metal catalyst. Citral is a polyfunctional molecule with a conjugated double bond, an isolated double bond and a carbonyl group. The conjugated double bond is much more reactive than the isolated one.

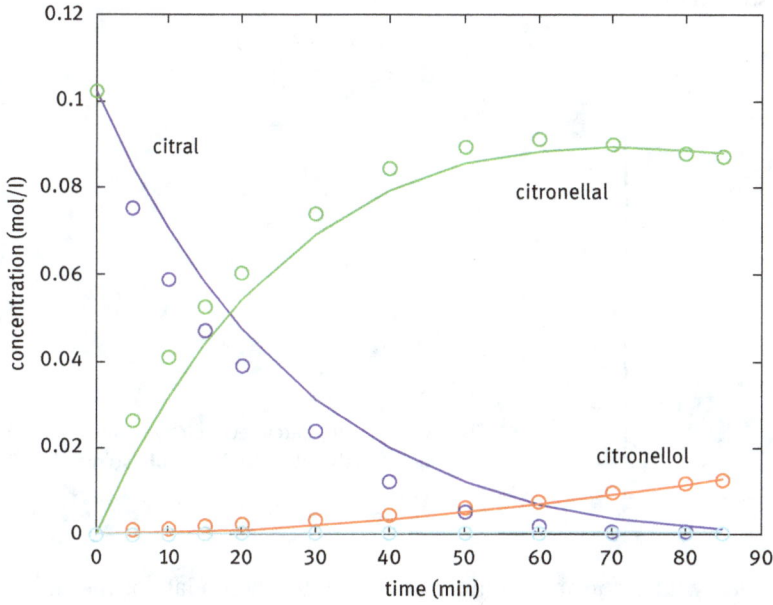

Figure 8.11: Typical kinetic behavior of a composite reaction system: hydrogenation of citral on a heterogeneous nickel catalyst (Salmi et al. 2007).

The first step is so rapid that it is difficult to obtain enough of samples during the first minutes of the experiment to record the kinetic behavior. Moreover, there is a risk that the gas–liquid mass transfer affects the overall kinetics of the rapid text. On the other hand, the third reaction step is very slow, which means that very long experiments are needed, and if the concentration of the final product is still low, it is impossible to estimate the numerical value of the last rate constant precisely. For the intermediate products (R and S), it is important to determine experimentally the concentration maxima in order to obtain good estimates of the rate constants. This cannot always be achieved with a standard isothermal experiment and a constant catalyst concentration.

Can this dilemma be surmounted? Is it possible to slow down the first step and accelerate the first step somehow? In case of heterogeneous catalysis, the observed rate is proportional to the catalyst bulk density (ρ_B). By starting with a large amount of liquid (V_L) the initial catalyst bulk density ($\rho_{0B} = m_{cat}/V_L$) is low and the first reaction is retarded. If liquid is continuously removed from the reactor during the kinetic experiment, the catalyst bulk density increases and the secondary and tertiary reactions are accelerated.

Let us assume a well-stirred slurry reactor, to which reactive gas is continuously added to compensate for the consumption of it, and liquid phase is pumped out with a constant rate. The solid catalyst remains inside the reactor vessel. The principle is illustrated in Figure 8.12.

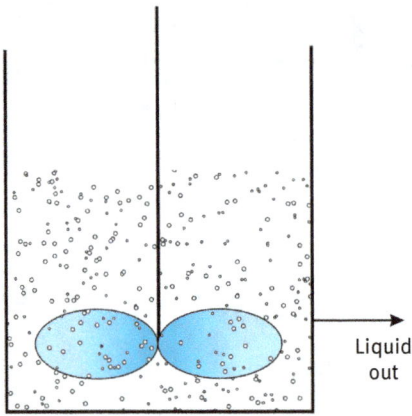

Figure 8.12: Semi-batch technology applied on heterogeneously catalysed liquid-phase reaction system.

If liquid is removed with a constant volumetric flow rate, the update of the liquid volume is

$$V = V_0 - V't \tag{8.93}$$

The catalyst bulk density is defined as

$$\rho_B = \frac{m_{cat}}{V_0 - V't} \tag{8.94}$$

By introducing the notations $\rho_{0B} = m_{cat}/V_0$ and V_0/V' and τ_0, we get

$$\rho_B = \frac{\rho_{0B}}{1 - t/\tau_0} \tag{8.95}$$

which clearly reveals the accelerating effect on the kinetics: the catalyst bulk density increases with time.

In the absence of external and internal mass transfer limitations, the mass balance of an arbitrary component in the liquid phase can be written as

$$r_i m_{cat} = \frac{dn_i}{dt} + V'c_i \tag{8.96}$$

The derivative dn_i/dt can be expressed with concentrations as follows,

$$dn_i/dt = d(c_iV)/dt \tag{8.97}$$

where $d(c_iV/dt)$ is

$$\frac{d(c_iV)}{dt} = V\frac{dc_i}{dt} + c_i\frac{dV}{dt} \tag{8.98}$$

Recalling that $dV/dt = -V'$, we obtain

$$\frac{dc_i}{dt} = \frac{r_i m_{cat}}{V_{OL} - V't} \tag{8.99}$$

which can easily be converted to the final form

$$\frac{dc_i}{dt} = \frac{\rho_{0B} r_i}{1 - t/\tau_0} \tag{8.100}$$

Example: first-order reaction

An irreversible first-order reaction

$$A \rightarrow P$$

is considered as an example. The rate equation $r_A = -kc_A$ is inserted in the mass balance giving

$$\frac{dc_A}{dt} = \frac{-k\rho_{0B} c_A}{1 - t/\tau_0} \tag{8.101}$$

This differential equation is easily solved by separation of the variables and integration within the limits [0,t] and [c0A, cA],

$$-\int \frac{dc_A}{c_A} = (k\rho_{0B}) \int \frac{dt}{1 - t/\tau_0} \tag{8.102}$$

The solution becomes

$$c_A/c_{0A} = (1 - t/\tau_0)^{k\rho_{0B}\tau_0} \tag{8.103}$$

A complete set of analytical solutions for the sequence $A \rightarrow R \rightarrow S$ is given in the reference (Salmi et al. 2007) and is not repeated here. The general approach is to solve the model eq. (8.100) numerically during the progress of the parameter estimation.

Thanks to this approach, the original kinetic pattern displayed in Figure 8.12 is transformed to a new one presented in Figure 8.13: the first reaction is slower than in the standard experiment (Figure 8.12), while the secondary and tertiary reactions are faster, which essentially improves the accuracy of the kinetic parameters.

The method can still be improved by applying temperature programming during the course of the experiment: if the temperature is increased during the experiment, the slow reactions are accelerated, as illustrated in Figure 8.14. In addition to

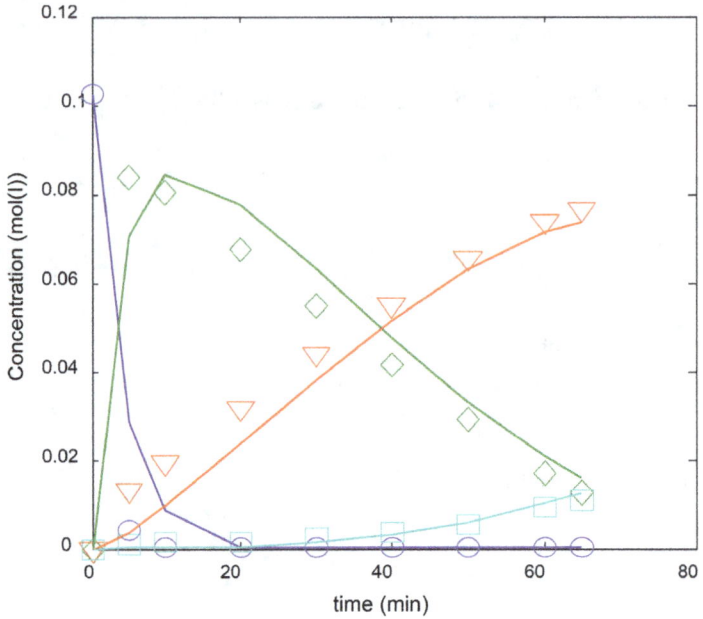

Figure 8.13: Transformed kinetic behavior of a composite reaction system: hydrogenation of citral on a heterogeneous nickel catalyst (Salmi et al. 2007).

temperature programming, liquid was continuously removed from the reactor during the experiment.

8.7.2 Composite reactions in the presence of a homogeneous catalyst

The approach presented in the previous section can be applied to homogeneously catalysed systems, too, but in an opposite way: if a solution of the homogeneous catalyst is continuously added to the reactor vessel, the catalyst concentration increases and the rates of the reactions are accelerated during the experiment. The principle is illustrated in Figure 8.15.

For a reactive component (i), the mass balance for the semi-batch reactor is

$$r_i V = \frac{dn_i}{dt} \qquad (8.104)$$

because the reactants and products are in batch. The derivative of the amount of substance is

8.7 Estimation of parameters from semi-batch experiments

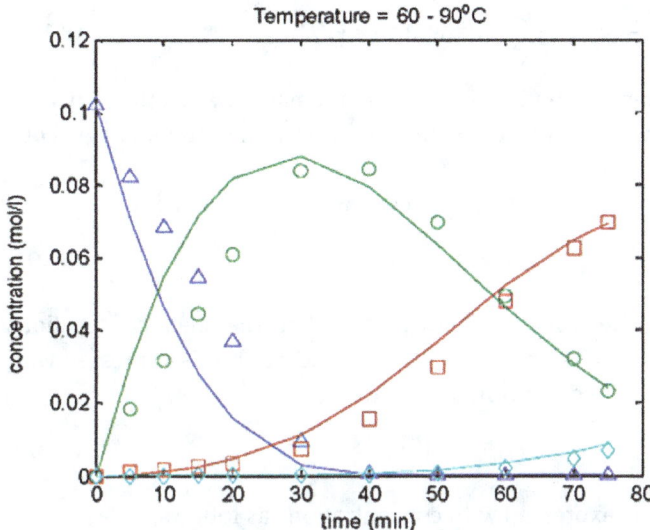

Figure 8.14: Non-isothermal semi-batch experiment for successful parameter estimation: hydrogenation of citral on a heterogeneous nickel catalyst (Salmi et al. 2007).

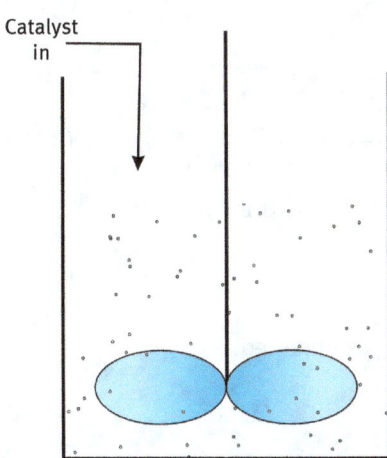

Figure 8.15: Semi-batch technology applied on homogeneously catalysed liquid-phase reaction system.

$$\frac{dn_i}{dt} = \frac{dc_i}{dt}V + c_i\frac{dV}{dt} \quad (8.105)$$

where dV/dt is equal to V' for a constant volumetric flow rate. After inserting this relation in the balance equation (8.104) and some rearrangement, the operative form of the balance equation is obtained,

$$\frac{dc_i}{dt} = r_i - \frac{c_i}{\tau_0 + t} \quad \text{where } \tau_0 = V_0/V' \tag{8.106}$$

The solution of the differential equation depends on the particular expression valid for the reaction kinetics and it is affected by the concentration of the homogeneous catalyst.

For the added homogeneous catalyst, the mass balance can be written as

$$n'_{0C} = dn_C/dt \tag{8.107}$$

because the catalyst is neither consumed nor produced in the chemical reactions going on in the reactor vessel. The inlet molar flow of the catalyst is expressed with the concentration in the feed and the volumetric flow rate,

$$n'_{0C} = c_{0C} V' \tag{8.108}$$

The derivative dn_C/dt can be expressed with concentrations as follows,

$$dn_C/dt = d(c_C V)/dt \tag{8.109}$$

where $d(c_C V/dt)$ is

$$\frac{d(c_C V)}{dt} = V \frac{dc_C}{dt} + c_C \frac{dV}{dt} \tag{8.110}$$

that is,

$$\frac{d(c_C V)}{dt} = (V_0 + V't) \frac{dc_C}{dt} + c_C V' \tag{8.111}$$

After inserting eqs. (8.108) and (8.111) in the mass balance (8.113), we obtain

$$c_{0C} V' = (V_0 + V't) \frac{dc_C}{dt} + c_C V' \tag{8.112}$$

By denoting $V_0/V' = \tau_0$ and rearranging, eq. (8.107) is obtained,

$$\frac{dc_C}{dt} = \frac{c_{0C} - c_C}{\tau_0 + t} \tag{8.113}$$

This ordinary differential equation (ODE) is easily solved by separation of variables and integration. The integration limits are $(0,t)$ and $(c_C(0), c_C)$. The solution is

$$c_C = \frac{c_{0C}(t/\tau_0) + c_C(0)}{1 + t/\tau_0} \tag{8.114}$$

In the beginning of the reaction $t = 0$ and $c_C = c_C(0)$ according to equation (8.114). On the other hand, if the catalyst addition is continued for a long time, $t/\tau_0 \to \infty$ and $c_C \to c_{0C}$, that is, the catalyst concentration in the feed.

The model consists of the mass balances of the reactive components, eq. (8.106) and the catalyst concentration according to eq. (8.114). For example, if a reaction A → P has the rate $R = kc_A c_C$, where A is the reactant, the generation rate is $r_A = -kc_A c_C$, which is inserted in the mass balance eq. (8.106). Analytical solutions can be derived for zero- and first-order reactions, but the general approach is the numerical solution of the balance equations during the parameter estimation.

Bibliography

Abramowitz, M. & Stegun I. A., Handbook of Mathematical Functions, Dover Publications Inc., New York, 1970.

Aris, R. (1975). The Mathematical Theory of Diffusion and Reaction in Permeable Catalysts. Oxford: Claredon Press.

Charpentier, J.-C. (1981). Mass-transfer rates in gas-liquid absorbers and reactors. In T. Drew, Advances in Chemical Engineering 11. New York: Academic Press. 69.

Danckwerts, P. (1970). Gas-Liquid Reactions. New York: McGraw-Hill.

Danckwerts, P. V. (1953). Continuous flow systems: Distribution of residence times. Chemical Engineering Science, 2(1), 1–13.

de Araujo Filho, C., Heredia, S., Eränen, K., & Salmi, T. (2016). Advanced millireactor technology for the kinetic investigation of very rapid reactions: Dehydrochlorination of 1,3-dichloro-2-propanol to epichlorohydrin. Chemical Engineering Science, 149(31), 35–41.

Deckwer, W. (1985). Reaktionstechnik in Blasensäulen. Aarau: Salle.

Fogg, P., & Gerrard, W. (1991). Solubility of Gases in Liquids. Chichester: Wiley.

Fott, P., & Schneider, P. (1984). Multicomponent Mass Transport with Complex Reactions in a Porous Catalyst. New Delhi: Wiley Eastern.

Froment, G., Bischoff, K., & de Wilde, J. (2011). Chemical Reactor Analysis and Design, 3rd Edition. Hoboken: Wiley.

Grénman, H., Salmi, T., Mäki-Arvela, P., Wärnå, J., Eränen, K., Tirronen, E., & Pehkonen, A. (2003). Modelling the kinetics of a reaction involving a sodium salt of 1,2,4-triazole and a complex substituted aliphatic halide. Organic Process Research & Development, 7, 942–950.

Higbie, R. (1935). Rate of absorption of a pure gas into still liquid during short periods of exposure. Transactions AIChe, 31, 365–389.

Hill, C. & Root C. (2014). An Introduction to Chemical Engineering Kinetics and Reactor Design, 2nd Edition. New York: Wiley.

Himmelblau, D., Jones, C., & Bischoff, K. (1967). Determination of rate constants for complex kinetics models. Industrial & Engineering Chemistry Fundamentals, 6(4), 539–543.

Hindmarsh, A. C. (1983). ODEPACK, a systematized collection of ODE solvers. In R. Stepleman, Scientific Computing 55–64. Amsterdam: IMACS/Noth Holland Publishing Company.

International Mathematical and Statistical Libraries (IMSL). (1987). Houston.

Kaps, P., & Wanner, G. (1981). A study of Rosenbrock-type methods of high order. Numerische Mathematik, 38, 279–298.

Keiski, R., Salmi, T., & Pohjola, V. (1992). Development and verification of a simulation model for a non-isothermal water-gas shift reactor. The Chemical Engineering Journal, 87(2), 17–29.

Laidler, K. (1987). Chemical Kinetics (3rd ed.). New York: Harper & Row.

Levenspiel, O. (1999). Chemical Reaction Engineering. New York: Wiley.

Lindfors, L. P., & Salmi, T. (1993). Kinetics of toluene hydrogenation on a supported Ni catalyst. Industrial and Engineering Chemistry Research, 32, 34–42.

Lindfors, L., Salmi, T., & Smeds, S. (1993). Kinetics of toluene hydrogenation on Ni/Al$_2$O$_3$ catalyst. Chemical Engineering Science, 48(22), 3813–3828.

Lundén, P (1991) ststionära simuleringsmodeller för katalytiska gasfasreaktioner i packade bäddreaktorer, Masters thesis, Laboratory of Industrial Chemistry and Reaction Engineerin, Åbo Akademi

Marquardt, D. (1963). An algorithm for least-squares estimation of nonlinear parameters. Journal of the Society for Industrial and Applied Mathematics, 11(2), 431–441.

Murzin, D. Y., & Salmi, T. (2016). Catalytic Kinetics. Amsterdam: Elsevier.

Perkins, L., & Geankoplis, C. (1969). Molecular diffusion in a ternary liquid system with the diffusion component dilute. Chemical Engineering Science, 24(7), 1035–1042.

Petzold, L. (1982). A description of DASSL: A differential/algebraic system solver. IMACS World Congress, 430–432.

Ramachandran, P., & Chaudhari, R. (1983). Three-Phase Catalytic Reactors. New York: Gordon and Breach Science Publishers.

Rase, F. H. (1977). Chemical Reactor Design for Process Plants. New York: Wiley.

Reid, R., Prausnitz, J., & Sherwood, T. (1988). The Properties of Gases and Liquids. New York: McGraw-Hill.

Romanainen, J. (1997). Numerical approach to modeling of dynamic bubble columns. Chemical Engineering and Processing, 36(1), 1–15.

Romanainen, J., & Salmi, T. (1992). The effect of reaction kinetics, mass transfer and flow pattern on non-catalytic and homogeneously catalyzed gas-liquid reactions in bubble columns. Chemical Engineering Science, 47(9–11), 2493–2498.

Romanainen, J., & Salmi, T. (1994). Numerical strategies in solving gas-liquid reactor models – 3. Steady state bubble columns. Computers & Chemical Engineering, 19(2), 139–154.

Russo, V., Kilpiö, T., Di Serio, M., Tesser, R., Santacesaria, E., Murzin, D., & Salmi, T. (2015). Dynamic non-isothermal trickle bed reactor with both internal diffusion and heat conduction: Sugar hydrogenation as a case study. Chemical Engineering Research and Design, 102, 171–185.

Salmi, T., Mäki-Arvela, P., Wärnå, J., Eränen, K., Denecheau, A., Alho, K., Murzin, D.Yu., Modelling of consecutive reactions with a semibatch liquid phase: Enhanced kinetic information by a new experimental concept, Industrial & Engineering Chemistry Research 46 (2007), 3912–3921.

Salmi, T., & Romanainen, J. (1995). A novel exit boundary condition for the axial dispersion model. Chemical Engineering and Processing: Process Intensification, 34(4), 359–366.

Salmi, T., & Wärnå, J. (1991). Modelling of catalytic packed-bed reactors-comparison of different diffusion models. Computers & Chemical Engineering, 15(10), 715–727.

Salmi, T., Mikkola, J.-P., & Wärnå, J. (2011). Chemical Reaction Engineering and Reaction Technology. Boca Raton: CRC Press.

Salmi, T., Wärnå, J. (1991). Modelling of catalytic packed bed reactors – comparison of different diffusion models. Computers & Chemical Engineering, 15(10), 715–727.

Scheibel, E. (1954). Correspondence. Liquid diffusivities. Viscosity of gases. Industrial & Engineering Chemistry Research, 46(9), 2007–2008.

Schiesser, W. (1991). The Numerical Method of Lines. San Diego: Academic Press.

Tirronen, E., & Salmi, T. (2003). Process development in the fine chemical industry. Chemical Engineering Journal, 91(2–3), 103–114.

Toppinen, S., Rantakylä, T.-K., Salmi, T., & Aittamaa, J. (1996). Kinetics of the liquid-phase hydrogenation of di- and tri-substituted alkylbenzenes over a nickel catalyst. Industrial & Engineering Chemistry Research, 35(12), 4424–4433.

Toppinen, S., Attamaa, J., Salmi, T. (1996). Interfacial mass transfer in trickle-bed reactor modelling. Chemical Engineering Science, 51(18), 4335–4345.

Toukoniitty, E., Wärnå, J., Murzin, D., & Salmi, T. (2010). Modelling of transient kinetics in catalytic three-phase reactors: Enantioselective hydrogenation. Chemical Engineering Science, 65, 1076–1087.

Trambouze, P., Landeghem, H., & Wauquier, J.-P. (1988). Chemical Reactors: Design, Engineering, Operation. Paris: Editions Technip.

van Santen, R. A., van Leeuwen, P. W., Moulijn, J. A., & Averill, B. (1999). Studies in Surface Science and Catalysis 123 – Catalysis: An Integrated Approach 123. Amsterdam, The Netherlands: Elsevier.

Vignes, A. (1966). Diffusion in binary solutions: Variation of diffusion coefficient with composition. Industrial & Engineering Chemistry Fundamentals, 5(2), 189–199.

Villadsen, J., & Michelsen, M. (1978). Solution of Differential Equation Models by Polynomial Approximation. Englewood Cliffs: Prentice-Hall.

Wakao, N. (1984). Particle-to-fluid heat/mass transfer coefficients in packed bed catalytic reactors. In L. Doraiswamy, Recent Advances in the Engineering analysis of Reacting Systems. New Delhi: Wiley Eastern, p. 14–37.

Wild, G., & Charpentier, J.-C. (1987). Diffusivité des gaz dans les liquids. Paris: Techniques d'ingenieur. 14–37.

Wilke, C., & Lee, C. (1955). Estimation of diffusion coefficients for gases and vapours. Industrial & Engineering Chemistry Research, 47(6), 1253–1257.

Exercises

E.1 Gas-phase tube reactor

Formic acid decomposes in a gas-phase system according to the reaction scheme displayed below:

$$HCOOH \rightarrow H_2O + CO \qquad (E.1.1)$$

$$HCOOH \rightarrow H_2 + CO_2 \qquad (E.1.2)$$

At 236 °C, the rate constants of reactions (E.1.1) and (E.1.2) are $k_1 = 2.8 \cdot 10^{-3}$ min^{-1} and $k_2 = 1.52 \cdot 10^{-4}$ min^{-1}, respectively. The reaction is conducted in an isothermal reactor with a constant total pressure of 1.0 bar. The reactor is fed with a mixture of inert gas and 50% vol. of formic acid. The inflow is 5 mol/h and the tube diameter is 5 cm.

a) Write the stoichiometric matrix for the system.
b) Define the molar balances for all components in a tubular reactor operating in accordance to plug flow conditions and axial dispersion.
c) Simulate the molar flows of all components as a function of the reactor length (1 meter). How long space-time is needed to convert 95% of formic acid?

E.2 Synthesis of maleic acid monoester in a semi-batch reactor

Maleic acid monoester (C) and water are formed as maleic acid (A) and hexanol (B) react in a batch reactor following the scheme below:

No solvent is present, and pure reactants are thus mixed together in a reaction vessel. The reaction mixture is heated until all A has melted at 53 °C. After this, the concentrations of A and B are as follows: $c_A^0 = 4.55\,\text{mol/L}$ and $c_B^0 = 5.24\,\text{mol/L}$. The reaction is of first order with respect to each reactant.

The temperature limit of 100 °C must not be exceeded due to risk of unwanted side reactions. On the other hand, the lowest possible temperature is the melting point of maleic acid 53 °C. A stirred batch reactor with a volume of 5.0 m³ is available. The overall heat transfer coefficient between the reactor and the surroundings can be approximated to U=250 W/m²/K. The conversion of maleic acid should reach a level of 95% as a minimum. The necessary data are displayed in Table E.2.1.

a) Introduce the mole and energy balances for the batch reactor.
b) Write a computer program for the simulation of the molar amounts and the temperature in the reactor.
c) Determine the heat transfer area so that the reaction time becomes as short as possible. How long reaction time is required in this case?
d) Could the batch reactor be operated adiabatically?

Table E.2.1: Data for maleic acid esterification.

Activation energy parameter	$E_a = 105\,\text{kJ/mol}$
Frequency factor	$A = 1.37 \cdot 10^{12}\,\text{L/mol/s}$
Reaction enthalpy	$\Delta H_r = -33.5\,\text{kJ/mol}$
Heat capacity	$\rho c_p = 1980\,\text{J/L/K}$
Total molar amounts	$c_A^0 = 4.55\,\text{mol/L},\ c_B^0 = 5.34\,\text{mol/L}$
Reactor volume	$V_R = 5\,\text{m}^3$
Overall heat transfer coefficient	$U = 250\,\text{W/m}^2/\text{K}$
Adiabatic temperature rise	$\Delta T_{ad} = 77\,\text{K}$

E.3 Exothermic reaction in a continuous stirred tank reactor

A chemical compound P will be produced in an adiabatic continuous stirred tank reactor (CSTR) via the homogeneous liquid phase reaction

$$A + B \rightarrow P$$

The kinetics of the reaction is described by the experimentally obtained second-order rate law:

$$R = k c_A c_B$$

a) Simulate the reactor dynamics, that is, the concentrations and the temperature in the reactor vessel, using suitable software. At the initial state the reactor is assumed to be filled with an inert solvent, the temperature of which can reside in the range of 290 ... 400 K.
b) Discuss the simulation results and plot the temperature and the concentrations versus the reaction time.
c) How many steady states are theoretically possible in this case?

The necessary data are given in Table E.3.1.

Table E.3.1: Data for second-order reaction in a CSTR.

Activation energy	$E_a = 99$ kJ/mol
Frequency factor	$A = 1.52 \cdot 10^9$ m^3/(mol·s)
Reaction enthalpy	$\Delta H_r = -33.5$ kJ/mol
Heat capacity	$\rho c_p = 1900$ kJ/m^{3}K
Feed concentrations	$c_{0A} = 5$ kmol/m^3, $c_{0B} = 5$ kmol/m^3
Feed temperature	$T_0 = 290$ K
Reactor volume	$V_R = 1.0$ m^3
Volumetric flow rate	$\dot{V} = 0.6 \cdot 10^{-3}$ m^3/s

E.4 Production of phtalic anhydride in a fixed bed reactor

The industrial production of phtalic anhydride is based on the oxidation of naphthalene or o-xylene. The reaction mechanism of the oxidation of o-xylene over a V_2O_5 –catalyst can in the presence of an excess of oxygen be simplified to a coupled consecutive and a parallel reaction

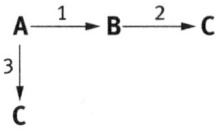

where A and B denote o-xylene and phtalic anhydride, respectively; C denotes the sum of the products of total oxidation (CO, CO_2, H_2O). The following rate equations are assumed to be valid:

$$R_1 = k_1 p_A p_O \quad (E.4.1)$$

$$R_2 = k_2 p_B p_O \quad (E.4.2)$$

$$R_3 = k_3 p_A p_O \quad (E.4.3)$$

where p_O denotes the partial pressure of oxygen. The oxidation reaction is carried out in a multitubular fixed bed reactor system, which is cooled with a molten salt bath. Due to the explosion risk, the concentration of o-xylene must be kept below 1 vol. % at the reactor inlet. The temperature is not allowed to exceed 660 K, since the catalyst is strongly deactivated at elevated temperatures exceeding the threshold value.

a) Describe the fixed bed with a two-dimensional pseudo-homogeneous model.
b) Compute and plot the temperature and phtalic acid concentration profiles as a function of the reactor length in a pilot plant reactor which consists of a single reactor tube and which operates under the conditions specified in Table E.4.1. Assume a one-dimensional pseudo-homogeneous model and consider four different inflow temperatures, $T_0 = 625, 630, 633$ and 635 K.
c) What is the maximum temperature in the reactor? In which way should the conditions be changed, if the maximum temperature is exceeded?
d) Which is the conversion of A and the selectivity towards B obtained in the reactor?
e) How would an increase in the temperature of the inflow affect the yield of B?

Table E.4.1: Data for xylene oxidation in a fixed bed reactor.

Activation energies	$E_{a1} = 1.133 \cdot 10^5$ J/mol
	$E_{a2} = 1.301 \cdot 10^5$ J/mol
	$E_{a3} = 1.200 \cdot 10^5$ J/mol
Frequency factors	$A_1 = 1.145 \cdot 10^8$ mol/(atm² · kg · s)
	$A_2 = 3.185 \cdot 10^8$ mol/(atm² · kg · s)
	$A_3 = 4.577 \cdot 10^7$ mol/(atm² · kg · s)
Reaction enthalpies	$\Delta H_1 = -1.2845 \cdot 10^6$ J/mol
	$\Delta H_2 = -3.276 \cdot 10^6$ J/mol
Initial mole fractions	$y_{0A} = 0.0093$ (xylene)
	$y_{0O} = 0.208$ (oxygen)
	$y_{0I} = 0.783$ (inert)
Specific heat capacity	$c_p = 1.046 \cdot 10^3$ J/(kg · K)
Temperature at the reactor inlet	$T_0 = 625, 630, 633, 635$ K
Total pressure at the inlet	$P_0 = 1$ atm
Molar mass at the inlet	$M = 29.48 \cdot 10^{-3}$ kg/mol
Mass flow	$\dot{m} = 6.386 \cdot 10^{-4}$ kg/s
Coolant temperature	$T_C = T_0$
Reactor length	$L = 3.00$ m
Reactor diameter	$d = 0.025$ m
Catalyst bulk density	$\rho_B = 1300$ Kg/m³
Overall heat transfer coefficient	$U = 96$ W/(m² · s · K)

E.5 Water–gas shift in a fixed bed reactor – diffusional limitations

The water–gas shift reaction is carried out in an isothermal packed bed

$$CO + H_2O \leftrightarrow CO_2 + H_2 \tag{E.5.1}$$

The bed is filled with iron-chromium oxide catalyst particles. The catalyst particles are cylindrical and have a diameter and a height of 3.2 mm. The temperature in the reactor can be set between 663 and 723 K, whereas the total pressure always remains 1 atm. The reaction rate (R') can be described by the equation:

$$R' = k \cdot c_{CO} \left(1 - \frac{c_{CO_2} c_{H_2}}{c_{CO} c_{H_2O}} \frac{1}{K} \right) \tag{E.5.2}$$

where K denotes the equilibrium constant for the reaction, and k is the first-order rate constant. The reaction is known to be strongly diffusion limited. Calculate the concentration profiles inside the catalyst pellet and the effectiveness factor at the inlet of the reactor tube at different operating temperatures. The kinetic and reactor data are summarized in Table E.5.1.

Compare your results to the analytical solution of the effectiveness factor for first-order reactions:

$$\eta_{eCO} = \left(\frac{3}{\varphi} \right) \left(\frac{1}{\tanh(\varphi)} - \frac{1}{\varphi} \right) \tag{E.5.3}$$

$$\varphi^2 = R_C^2 \left(\frac{k \cdot \rho_P}{D_{eCO}} \right) \tag{E.5.4}$$

Keep in mind the following relations when solving the problem:

$$D_{ei} = \left(\frac{\varepsilon_P}{\tau_P} \right) \left(\frac{1}{D_{mi}} + \frac{1}{D_{Ki}} \right)^{-1} \tag{E.5.5}$$

$$\frac{D_{Ki}(T_1)}{D_{Ki}(T_2)} = \sqrt{\frac{T_1}{T_2}} \tag{E.5.6}$$

$$\frac{D_{Ki}(T_1)}{D_{Kj}(T_1)} = \sqrt{\frac{M_j}{M_i}} \tag{E.5.7}$$

$$\eta_{ei} = \frac{(s+1) \int_0^{R_C} r_i \cdot r^s dr}{r_i(c^b) R_C^{s+1}} \tag{E.5.8}$$

Table E.5.1: Data for water–gas shift reaction. The Knudsen diffusion coefficient and rate constant were determined by Keiski et al. (1992).

First-order rate constant	$k = A_{+1} \exp\left(-\frac{E_A}{RT}\right)$
Activation energy	$E_A = 94.3$ kJ/mol
Frequency factor	$A_{+1} = 95.02$ m^3/g/s
Catalyst properties	$\rho_B = 1.55$ g/cm^3
	$\varepsilon_P = 0.55$
	$\tau_P = 2$
	$R_C = 3.2$ mm
	$s = 2$
Reactor data	$T = 663, 683, 703, 723$ K
	$P = 1$ atm
Inlet molar flow composition	$y_{0,CO} = 0.07$
	$y_{0,H_2O} = 0.5$
	$y_{0,CO_2} = 0.03$
	$y_{0,H_2} = 0.2$
	$y_{0,N_2} = 0.2$
CO Knudsen diffusivity @ 723 K	$D_{K,CO} = 0.107$ cm^2/s
Equilibrium constant	$K = \exp\left(\frac{4577.8}{T} - 4.33\right)$
Gas–solid mass transfer coefficient	Very large

E.6 Steady-state CSTR's in series: oxidation of Iron (II) to Iron(III)

Ferric ions, that is, Fe^{3+} or Fe(III), are an important coagulant for water treatment process. They are produced by the oxidation of iron sulphate, $FeSO_4$, in acidic medium and in the presence of air. Naturally, water is the solvent for this reaction. The overall oxidation process is

$$Fe^{2+} + \frac{1}{4}O_2 + H_3O^+ \rightarrow Fe^{3+} + \frac{3}{2}H_2O$$

The reaction is usually conducted catalytically by the addition of active carbon, however, the homogeneous reaction is quite significant and cannot be neglected. The homogeneous and heterogeneous reaction rates are presented by eqs. (E.6.1, E.6.2):

$$r_{hom} = \frac{A_1 c_{Fe^{2+}}^2 c_{O_2}}{1 + \frac{A_1}{A_2} c_{Fe^{2+}}} \exp\left[-\frac{Ea_{hom}}{R}\left(\frac{1}{T} - \frac{1}{T_{ref}}\right)\right] \tag{E.6.1}$$

$$r_{het} = A_3 \exp\left[-\frac{Ea_{het}}{R}\left(\frac{1}{T} - \frac{1}{T_{ref}}\right)\right] c_{Fe^{2+}} \sqrt{c_{O_2}} \tag{E.6.2}$$

The oxidation of iron sulphate is conducted in two isothermal isobaric CSTRs connected in series. The temperature of the process is 100 °C and total pressure of air, in both reactors, is 3.5 atm. The concentration of the iron sulphate in the inlet feed of the first reactor is 2.0 mol/L and the liquid volume in each reactor is 1 m³. The gas–liquid mass transfer resistance is negligible, therefore, the concentration of oxygen in the liquid phase can be directly estimated by the Henry's law:

$$c_{L,O_2} = \frac{P_{O_2} c_L}{H} \tag{E.6.3}$$

where P_{O_2} is the partial pressure of oxygen in the gas phase, c_L is the total concentration of compounds in liquid phase and H is the Henry constant. The Henry constant depends on the composition of the liquid-phase in the following way:

$$H(atm) = \frac{10^{\sum c_i [h_i + h_T(T - 298)]}}{\exp\left(A + \frac{B}{T} + C \cdot \ln(T)\right)} \tag{E.6.4}$$

where T is the temperature in Kelvin. The aim of the process is to achieve a global conversion of 95% of ferrous ions (Fe^{2+}) with a space-time no longer than 500 minutes in each reactor.

a) Derive the general equations that describe the concentrations of ferric and ferrous ions in both reactors operating at steady-state conditions as a function of the space-time.

b) Would it be possible to meet the required conditions by operating both reactors without any addition of catalyst? If so, at which global space-time is the 95% conversion achieved?

c) Assume that both reactors operate with a catalyst load of 120 kg/m³, would the process requirements be satisfied? If so, at which global space-time is the 95% conversion achieved?

d) Consider that the first reactor operates without the addition of catalyst and the second one operates with 120 kg/m³ of catalyst loading. Is the global conversion requirement achieved? If so, at which global space-time does it occur?

e) Discuss the results obtained in items b), c) and d). Which design would you choose to operate? Why?

f) BONUS! If you were to design a similar system for a water treatment plant, how would you adjust the parameters to maximize the conversion of Fe^{2+}? Consider that the reactors may have different sizes and operate at different conditions. Discuss with your colleagues the pros and cons of changing each parameter (temperature, pressure, catalyst loading, reactor volume) and if these changes are feasible in practice. Then ask yourselves: is there another reactor design which would suit this application better?

Table E.6.1: Reaction and reactor data.

Reference temperature	$T_{ref} = 353.15$ K
Homogeneous reaction kinetic parameters	$A_1 = 21.6$ L² · mol⁻² · min⁻¹
	$A_2 = 36.8$ L · mol⁻¹ · min⁻¹
	$\frac{Ea_{hom}}{R} = 4176$ K
Heterogeneous reaction kinetic parameters	$A_3 = 5.07$ L$^{3/2}$ · mol$^{-1/2}$ · kg$_{cat}^{-1}$ min⁻¹
	$\frac{Ea_{het}}{R} = 3930$ K
Henry constant parameters	$A = -171.2542$
	$B = 8391.24$
	$C = 23.24323$
	$h_{H^+} = 0$
	$h_{Fe^{2+}} = 0.1523$ L · mol⁻¹
	$h_{Fe^{3+}} = 0.1161$ L · mol⁻¹
	$h_{SO_4^{-2}} = 0.1117$ L · mol⁻¹
	$h_T = 3.34 \cdot 10^{-4}$ L · mol⁻¹
Reactors conditions	$P = 3.5$ atm
	$T = 100$ °C
	$V_1 = V_2 = 1$ m³
Concentration of FeSO₄ in the inlet feed	$[c_{FeSO_4}]_0 = 2.0$ mol/L

E.7 A fluidized bed reactor

A first-order, irreversible catalytic gas-phase reaction

$$A \rightarrow P$$

should be carried out in an isothermal fluidized bed.
a) Simulate the conversion of A as a function of the bed length coordinate.
b) What is the bed height for a conversion level of 95%?

Table E.7.1: Data.

ρ_{Bb}	7.5 kg/m³
$\rho_{Bc} = V_c/V_b$	290 kg/m³
$\rho_{Be} = V_e/V_b$	1020 kg/m³
K_{bc}	1.4 s⁻¹
K_{ce}	0.9 s⁻¹
k	1.5 m³/(kg·h)
w	1800 m/h
w_{mf}	20.5 m/h
d_b	0.1 m

E.8 Three-phase slurry reactor: Hydrogenation of aromatics

Catalytic hydrogenation of an aromatic compound proceeds on the surface of a Ni/Al$_2$O$_3$ catalyst according to the stoichiometry

$$T + 3H_2 \rightarrow TH$$

The reaction rate is defined as

$$R = \frac{kK_T K_H c_T c_H}{\left(1 + 3K_T c_T + \sqrt{K_H c_H}\right)^3}$$

The hydrogenation takes place in an isothermal semi-batch reactor at 373 K and at a hydrogen pressure of 20 bar. The reaction starts with pure aromatic compound in the reactor. The reactor volume is 1.1 L and the liquid volume in the reactor is 1.0 L. The initial concentration of reactant is 9.5 mol/L and the organic compounds are assumed to stay in the liquid phase at the prevailing conditions. The data needed as input are summarized in Table E.8.1. Liquid-solid mass transfer resistance is negligible for all the following items.

a) What is the reaction time required to achieve 99% conversion in case there are no gas–liquid mass transfer limitations and no diffusion limitations inside the particle?
b) Consider that the stirring in the reactor is actually not so efficient. What is the required reaction time needed to achieve 99% conversion if the gas–liquid mass transfer parameter for hydrogen is $k_{L,H_2} \cdot a = 0.05$ s^{-1} and no internal diffusion limitations are assumed?
c) How long reaction time is needed to achieve 99% conversion in case the mass transfer inside the particle is taken into account in item b)?
d) Now, assume that the external mass transfer limitation is negligible and compute the required reaction time to achieve 99% conversion considering internal mass transfer limitations.
e) Plot the conversion vs. time for items a), b), c) and d) and discuss the obtained results. How would you distinguish the effects of external and internal mass transfer limitations experimentally?

Table E.8.1: Reaction, reactor and catalyst data.

Total pressure	$P_0 = 20$ bar
Reaction temperature	$T_0 = 373.2$ K
Reactor volume	$V_R = 1.1$ L
Liquid volume	$V_L = 1.0$ L
Reaction rate constant	$k(T_0) = 2.1$ mol/kg/s
Aromatic adsorption constant	$K_T(T_0) = 0.25$ L/mol
Hydrogen adsorption constant	$K_H(T_0) = 37.0$ L/mol
Aromatic initial concentration	$c_T^0 = 9.5$ mol/L
H_2 mole fraction at saturation	$x_{H_2}^*(T_0) = 0.014$
Gas–liquid mass transfer parameter for H_2	$k_{L,H_2} \cdot a = 0.05$ s^{-1}
Particle radius	$R_C = 0.2$ mm
Particle porosity	$\varepsilon_P = 0.5$
Particle tortuosity	$\tau = 4$
Particle density	$\rho_P = 1300$ kg/m^3
Catalyst mass	$m_{cat} = 40$ g
Hydrogen diffusivity	$D_{H_2} = 4.4 \cdot 10^{-9}$ m^2/s
Organic compounds diffusivities	$D_T \approx D_{TH} = 5.0 \cdot 10^{-11}$ m^2/s

*Exercise inspired by the work of Toppinen et al., *Ind. Eng. Chem. Res.*, 35, 1824–1833, 1996.

E.9 Chlorination of p-cresol in a continuous stirred tank reactor

Chlorination of *para*-cresol occurs in two consecutive steps according to the reaction scheme

p-cresol + Cl–Cl $\xrightarrow{k_1}$ mono-chloro-p-cresol + HCl   (E.9.1)

mono-chloro-p-cresol + Cl–Cl $\xrightarrow{k_2}$ di-chloro-p-cresol + HCl   (E.9.2)

The reaction will be carried out in an inert solvent, carbon tetrachloride (CCl_4). The liquid which flows through the reactor contains initially 0.2 mol/dm³ *para*-cresol. The gas flow of (Cl_2) through the reactor is assumed to be so high that the liquid phase is saturated with respect to chlorine. The gas flow contains 20% chlorine. The saturation concentration of chlorine was determined to amount to 0.44 mol/dm³ at the chlorine partial pressure of 0.2 atm and at 0 °C.

The rate constant of reactions (E.9.1) and (E.9.2) have the following values at 0 °C:

$$k_1 = 5.620 \, dm^3/mol \cdot s$$

$$k_2 = 0.107 \, dm^3/mol \cdot s$$

The chlorination process will be carried out in a continuous stirred tank reactor which operates isothermally at 0 °C and at atmospheric pressure.

Assume elementary reaction rates for reactions (E.9.1) and (E.9.2). Determine the liquid space-time that gives the highest yield of the mono-chlorinated product. Plot the concentration profiles as a function of the liquid-phase residence time.

E.10 Reaction between methanol and triphenyl methyl chloride

Initially, 0.054 mol/L methanol reacts with 0.106 mol/L triphenyl methyl chloride in dry benzene solution.

$$CH_3OH + (C_6H_5)CCl \rightarrow (C_6H_5)COCH_3 + HCl$$

Determine the rate law and the rate constants from the enclosed data (Table E.10.1).

Table E.10.1: Kinetic data as a function of time.

t/min	426	1150	1660	3120
c_{CH_3OH}/(mol/L)	0.0351	0.0222	0.0186	0.0124

E.11 Use of millireactor for the kinetic study of very fast reaction: Dehydrochlorination of 1,3-dichloro-2-propanol

The dehydrochlorination of 1,3-dichloro-2-propanol (αγ-DCP) with an inorganic base produces 1-chloro-2,3-epoxypropane (epichlorohydrin), which is an important building block for the production of epoxy resins and plasticizers. The overall dehydrochlorination reaction is depicted in Figure E.11.1.

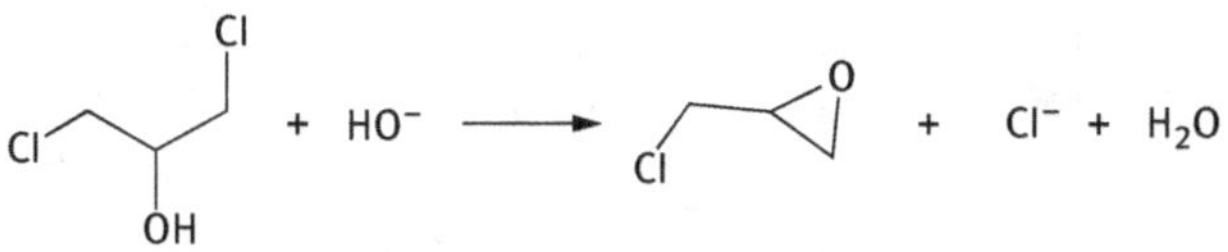

Figure E.11.1: Overall reaction scheme of the dehydrochlorination of αγ-DCP.

The reaction between αγ-DCP and sodium hydroxide is very rapid, which makes it challenging to study the dehydrochlorination kinetics in classical batch or tubular reactors. For this reason, de Araujo et al. (2016) decided to employ a tubular isothermal millireactor, which allowed samples to be withdrawn along the reactor length, at residence times as small as 1 sec. After stablishing careful reaction and sampling procedures, experimental data were obtained for very diluted aqueous solutions containing equimolar concentrations of αγ-DCP and NaOH at different temperatures. The data are presented in Table E.11.1. Assuming that the effects of axial dispersion are negligible in the system,

a) List few advantages and disadvantages of the use of micro/milli scale reactors.
b) Determine the rate constant, at each temperature, assuming first order kinetics with respect to αγ-DCP and zero order with respect to sodium hydroxide. Make an Arrhenius plot for determining the activation energy and pre-exponential factor for this kinetic model.
c) Determine the rate constant, at each temperature, assuming first order kinetics with respect to both αγ-DCP and NaOH. Make an Arrhenius plot for determining the activation energy and pre-exponential factor for this kinetic model.
d) Based on the data obtained in b) and c), which one is the most appropriate kinetic model for describing the experimental data?
e) Now that you have chosen the kinetic model, determine the activation energy and the pre-exponential factor from the data without making an Arrhenius plot. NOTE: in this case, you need to use equation (8.49).
f) Make a 3D sensitivity plot of the parameters estimated in item e).
g) BONUS! Think about a condensed reaction mechanism that would describe the dehydrochlorination reaction showed in Figure E.11.1. Derive the dehydrochlorination reaction rate and check if it coincides with the ones discussed in items a) or b).

Table E.11.1: Experimental data obtained for the dehydrochlorination in millireactor.

T=30 °C		T=40 °C	
Residence Time (s)	αγ-DCP (mmol/L)	Residence Time (s)	αγ-DCP (mmol/L)
0.000	10.705	0.000	9.295
0.982	10.525	0.982	8.595
1.178	10.475	1.178	8.195
3.191	10.265	3.191	7.560
3.829	10.150	3.829	6.970
7.118	9.220	7.118	6.000
8.541	9.100	8.541	5.480
18.467	7.160	18.467	3.415
22.160	7.005	22.160	3.205
T=50 °C		T=60 °C	
Residence Time (s)	αγ-DCP (mmol/L)	Residence Time (s)	αγ-DCP (mmol/L)
0.000	8.605	0.000	6.080
0.982	7.185	0.982	4.825
1.178	6.545	1.178	4.030
3.191	5.745	3.191	3.130
3.829	5.130	3.829	2.935
7.118	4.120	7.118	1.860
8.541	3.360	8.541	1.660
18.467	1.990	18.467	0.930
22.160	1.795	22.160	0.810

NOTE! For the sake of simplicity, the solution of this exercise may be accomplished by analysing the data of the decay of αγ-DCP along the reactor length. Remember that sodium hydroxide and αγ-DCP are present at equimolar concentrations. Due to the design characteristics of the millireactor, a certain amount of αγ-DCP had already reacted at the residence time "zero." Nevertheless, this should not disturb the parameter estimation. For further details about the reactor design and the dehydrochlorination reaction, please check de Araujo Filho et al., *Chem. Eng. Sci.*, 149 (31), 35–41, 2016.

E.12 Multiple liquid-phase reaction system

Grénman et al. studied the reactions of triazole in a liquid-phase batch reactor

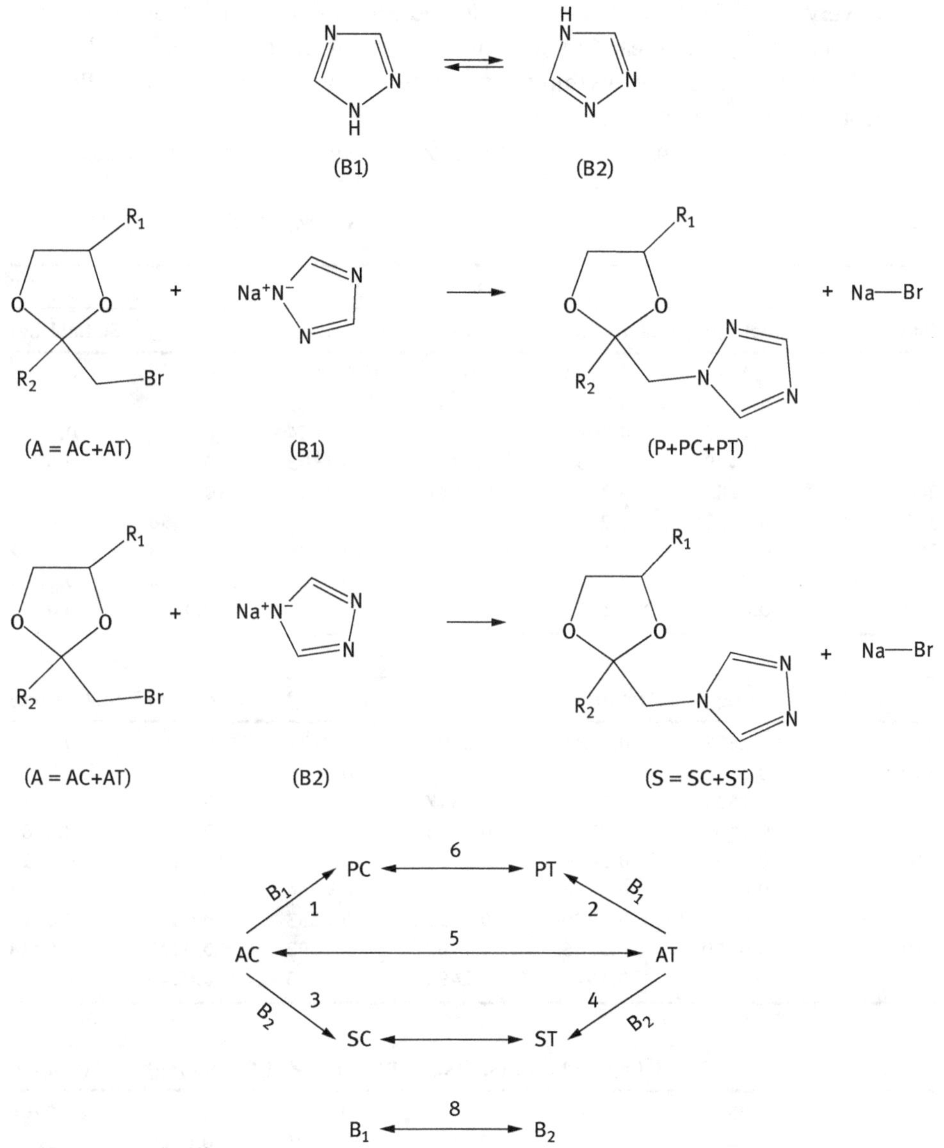

Figure E.12.1: The reaction schemes. The desired product, PC and PT, are formed when the triazole B_1 attaches to the reactants AC and AT. The unwanted side products are SC and ST and are formed when triazole in the form B_2 reacts.

a) Investigate the system thoroughly and reveal which constants/merged constants can be estimated preliminary by means of linear regression assuming that all reactions are elementary.
b) Experimental observations suggested that the isomerization reactions 5, 6 and 7 are very slow. On the other hand, the equilibrium step 8 is stablished at the very beginning of the reaction. Rewrite the mass balance equations for this particular case. Estimate everything you can by simple linear regression. Prepare illustrative linear plots.
c) Estimate the final values of the parameters by means of nonlinear regression.

Table E.12.1: Kinetic data from H. Grénman et al., *Org. Proc. Res. Dev.*, 7 (6), 942–950, 2003.

	120 °C					
Time (min)	AT (mol/kg)	AC (mol/kg)	PT (mol/kg)	PC (mol/kg)	ST (mol/kg)	SC (mol/kg)
0	0.2685	0.3993	0.0194	0.0171	0	0
60	0.213	0.3471	0.0682	0.0621	0	0
120	0.1716	0.3032	0.102	0.0964	0.008	0.0099
180	0.1439	0.2752	0.1334	0.1303	0.0112	0.0144
240	0.1182	0.2429	0.1531	0.1537	0.0137	0.0177
300	0.0998	0.2199	0.1719	0.177	0.0158	0.0208
360	0.083	0.1967	0.1872	0.1978	0.0178	0.0239
420	0.0704	0.1797	0.2052	0.2222	0.0194	0.0268
480	0.0574	0.1581	0.2137	0.2368	0.0209	0.0293

	125 °C					
Time (min)	AT (mol/kg)	AC (mol/kg)	PT (mol/kg)	PC (mol/kg)	ST (mol/kg)	SC (mol/kg)
0	0.2673	0.3981	0.0195	0.0172	0	0
60	0.198	0.3347	0.0862	0.08	0	0
120	0.1504	0.2851	0.1317	0.1282	0	0.023
180	0.1105	0.2337	0.1583	0.1609	0.0134	0.018
240	0.0846	0.2006	0.1906	0.2016	0.0172	0.0237
300	0.066	0.1744	0.2139	0.2344	0.0202	0.0283
360	0.0487	0.1439	0.2255	0.2553	0.0218	0.0315
420	0.0359	0.1186	0.2309	0.2693	0.0229	0.0338
480	0.0281	0.104	0.2493	0.2983	0.0246	0.0373

	130 °C					
Time (min)	AT (mol/kg)	AC (mol/kg)	PT (mol/kg)	PC (mol/kg)	ST (mol/kg)	SC (mol/kg)
0	0.2377	0.3684	0.0441	0.0406	0	0.0264
60	0.1492	0.2762	0.1145	0.1104	0.0209	0.0242
120	0.1012	0.2216	0.1659	0.1708	0.026	0.032
180	0.064	0.1659	0.1884	0.2054	0.023	0.0299
240	0.0424	0.1302	0.2143	0.2453	0.0241	0.0329
300	0.0272	0.0992	0.2266	0.2702	0.0269	0.0384

Table E.12.1 (continued)

			130 °C			
Time (min)	AT (mol/kg)	AC (mol/kg)	PT (mol/kg)	PC (mol/kg)	ST (mol/kg)	SC (mol/kg)
360	0.0177	0.0767	0.2369	0.2924	0.0288	0.0426
420	0.0113	0.0577	0.2377	0.3015	0.0288	0.0437
480	0.0074	0.0449	0.2477	0.3215	0.0301	0.0468
			135 °C			
Time (min)	AT (mol/kg)	AC (mol/kg)	PT (mol/kg)	PC (mol/kg)	ST (mol/kg)	SC (mol/kg)
0	0.2073	0.3419	0.0704	0.0663	0.0076	0.0099
60	0.1091	0.2348	0.1541	0.1569	0.0173	0.0211
120	0.0595	0.1621	0.1982	0.2178	0.0232	0.0305
180	0.0297	0.106	0.2147	0.253	0.0267	0.0378
240	0.0161	0.0734	0.2287	0.2829	0.0287	0.0428
300	0.0087	0.0509	0.2486	0.3188	0.0326	0.0505
360	0	0.0387	0.2482	0.3276	0.03	0.0479
420	0	0.0257	0.25	0.3368	0.0329	0.0535
480	0	0.0171	0.2499	0.3416	0.031	0.0513
			142 °C			
Time (min)	AT (mol/kg)	AC (mol/kg)	PT (mol/kg)	PC (mol/kg)	ST (mol/kg)	SC (mol/kg)
0	0.2412	0.375	0.0433	0.0391	0	0
60	0.0884	0.204	0.1797	0.1899	0.0205	0.0268
120	0.0312	0.1061	0.2204	0.2603	0.0262	0.038
180	0.0121	0.0612	0.2575	0.3273	0.0315	0.0489
240	0	0.035	0.2549	0.3395	0.0311	0.0506
300	0	0.0195	0.2581	0.3521	0.0321	0.0534
360	0	0.0103	0.2572	0.357	0.0316	0.0533
420	0	0	0.2676	0.3747	0.033	0.0563
480	0	0	0.2562	0.3602	0.0316	0.054
			147 °C			
Time (min)	AT (mol/kg)	AC (mol/kg)	PT (mol/kg)	PC (mol/kg)	ST (mol/kg)	SC (mol/kg)
0	0.2346	0.3903	0.0896	0.0834	0.0189	0.03
60	0.0582	0.1683	0.2485	0.2817	0.0308	0.0421
120	0.012	0.061	0.2582	0.3284	0.0339	0.0515
180	0	0.0258	0.259	0.3495	0.0344	0.0555
240	0	0.0104	0.2781	0.3852	0.0377	0.0624
300	0	0	0.2666	0.3734	0.0354	0.0592
360	0	0	0.2548	0.3581	0.0339	0.0569
420	0	0	0.2973	0.4181	0.0403	0.0676
480	0	0	0.3036	0.427	0.0428	0.0718

Table E.12.1 (continued)

	150 °C					
Time (min)	AT (mol/kg)	AC (mol/kg)	PT (mol/kg)	PC (mol/kg)	ST (mol/kg)	SC (mol/kg)
0	0.2033	0.3348	0.0714	0.0661	0.0078	0.0092
60	0.0376	0.1187	0.2157	0.2507	0.0277	0.0392
120	0	0.0456	0.253	0.3323	0.0334	0.0533
180	0	0.0177	0.2529	0.3465	0.0337	0.0558
240	0	0	0.256	0.3578	0.034	0.0575
300	0	0	0.261	0.3666	0.0345	0.0588
360	0	0	0.2547	0.3586	0.0339	0.0575
420	0	0	0.2563	0.3616	0.0342	0.0581
480	0	0	0.2598	0.3665	0.0346	0.0587

E.13 Gas–liquid reactions in a semi-batch reactor

Butanoic acid is chlorinated in the presence of a homogeneous catalyst, chlorosulphonic acid, to α-monochloro butanoic acid according to reaction (E.13.1):

$$\text{(CA)} + Cl-Cl \longrightarrow \text{(MCA)} + HCl \qquad (E.13.1)$$

The chlorination experiments were conducted in a semi-batch laboratory scale reactor by bubbling gaseous chlorine through the liquid batch, which initially consisted of pure butanoic acid and the homogeneous catalyst. The chlorine concentration in the liquid phase was constant during the course of the reaction. The reaction rates of (E.13.1) is given by the expression:

$$\frac{dx_{MCA}}{dt} = k' + k_1 x_{MCA}^{1/2} \qquad (E.13.2)$$

where k' and k_1 are constants and x_{MCA} is the mole fraction of MCA in the liquid phase. The catalyst concentration is included in k_1 and k'. In the experiments, the mole fractions of MCA were registered as a function of the reaction time. The results are listed in Table E.13.1.

Estimate the constants k' and k_1 from the experimental data using regression analysis. Estimate also the temperature dependence of the constants k' and k_1 using the Arrhenius' law.

Table E.13.1: Experimental data from the chlorination of butanoic acid.

T = 90 °C		T = 110 °C	
t/min	x_{MCA}	t/min	x_{MCA}
0	0	0	0
20.0	0.0000	20.0	0.0131
40.0	0.0119	40.0	0.0386
60.0	0.0264	60.0	0.0929
80.0	0.0567	80.0	0.1780
100.0	0.1113	100.0	0.2769
120.0	0.1556	120.0	0.3876
T = 120 °C		T = 130 °C	
t/min	x_{MCA}	t/min	x_{MCA}
0	0	0	0
20.0	0.0153	20.0	0.0493
40.0	0.0543	40.0	0.1199
60.0	0.1934	60.0	0.2648
80.0	0.3099	80.0	0.4235
100.0	0.4594	100.0	0.5875
120.0	0.5901	120.0	0.7442

E.14 Gas-phase reaction in a differential reactor

Hydrogenation of toluene to methylcyclohexane is an important model reaction for dearomatization processe

$$\text{Toluene (T)} + 3\,H_2 \longrightarrow \text{Methylcyclohexane (MCH)} \qquad (E.14.1)$$

The hydrogenation reaction is carried out on a supported Ni-catalyst. Kinetic experiments were performed out in a laboratory scale differential reactor operating at atmospheric pressure and at constant temperatures. The experimental data are shown in Table E.14.1. In the present case, the hydrogenation process can be regarded as an irreversible reaction, and three different kinetic equations have been proposed for the forward reaction:

$$R_1 = k \cdot p_H^m \cdot p_T^n \qquad (E.14.2)$$

$$R_2 = \frac{k K_T K_H^3 p_T p_H^3}{\left(1 + K_T p_T + (K_H p_H)^{1/2}\right)^7} \qquad (E.14.3)$$

$$R_2 = \frac{k K_T K_H p_T p_H}{\left(1 + 3 K_T p_T + (K_H p_H)^{1/2}\right)^3} \qquad (E.14.4)$$

where k is the rate parameter, and K_H and K_T denote the adsorption parameters of hydrogen and methylcyclohexane, respectively. The temperature dependence of the reaction parameter is given by

$$k = A \cdot e^{\frac{-E_a}{RT}} \qquad (E.14.5)$$

where A and E_a denote the frequency factor and the activation energy, respectively; the adsorptions terms may be assumed constant.
a) Which are the basic parameters that can be estimated from the experimental data?
b) Estimate the parameter values by regression analysis.
c) Which of the models is the best one?
d) Is it possible to estimate the temperature dependences of the adsorption constants from the kinetic data provided in Table E.14.1? Comment.

Table E.14.1: Experimentally measured reaction rates on toluene hydrogenation.

	T = 120 °C				T = 135 °C		
N^0	p_{H_2} (atm)	p_T (atm)	$10^6 \cdot R$ (mol/(kg·s))	N^0	p_{H_2} (atm)	p_T (atm)	$10^6 \cdot R$ (mol/(kg·s))
1	0.808	0.125	0.309	1	0.808	0.125	0.683
2	0.808	0.125	0.276	2	0.808	0.125	0.600
3	0.605	0.125	0.207	3	0.605	0.125	0.572
4	0.605	0.125	0.280	4	0.605	0.125	0.549
5	0.398	0.125	0.195	5	0.398	0.125	0.441
6	0.398	0.125	0.163	6	0.398	0.125	0.433
7	0.607	0.180	0.137	7	0.607	0.180	0.335
8	0.607	0.180	0.131	8	0.607	0.180	0.364
9	0.605	0.214	0.120	9	0.605	0.214	0.285
10	0.605	0.214	0.100	10	0.605	0.214	0.228
11	0.600	0.266	0.075	11	0.600	0.266	0.173
12	0.600	0.266	0.068	12	0.600	0.266	0.188
13	0.609	0.323	0.059	13	0.609	0.323	0.119
14	0.609	0.323	0.057	14	0.609	0.323	0.141

	T = 151 °C				T = 171 °C		
N^0	p_{H_2} (atm)	p_T (atm)	$10^6 \cdot R$ (mol/(kg·s))	N^0	p_{H_2} (atm)	p_T (atm)	$10^6 \cdot R$ (mol/(kg·s))
1	0.808	0.125	1.813	1	0.808	0.125	4.265
2	0.808	0.125	1.653	2	0.808	0.125	3.638
3	0.605	0.125	1.395	3	0.605	0.125	3.261
4	0.605	0.125	1.499	4	0.605	0.125	3.363
5	0.398	0.125	0.893	5	0.398	0.125	2.301
6	0.398	0.125	0.931	6	0.398	0.125	2.672
7	0.607	0.180	0.831	7	0.607	0.180	1.692
8	0.607	0.180	0.815	8	0.607	0.180	2.007
9	0.605	0.214	0.485	9	0.605	0.214	1.293
10	0.605	0.214	0.610	10	0.605	0.214	1.545
11	0.600	0.266	0.386	11	0.600	0.266	0.893
12	0.600	0.266	0.347	12	0.600	0.266	0.893
13	0.609	0.323	0.277	13	0.609	0.323	0.632
14	0.609	0.323	0.260	14	0.609	0.323	0.833

	T = 186 °C				T = 200 °C		
N^0	p_{H_2} (atm)	p_T (atm)	$10^6 \cdot R$ (mol/(kg·s))	N^0	p_{H_2} (atm)	p_T (atm)	$10^6 \cdot R$ (mol/(kg·s))
1	0.808	0.125	8.281	1	0.808	0.125	15.681
2	0.808	0.125	7.469	2	0.808	0.125	13.939
3	0.605	0.125	6.133	3	0.605	0.125	11.205
4	0.605	0.125	7.123	4	0.605	0.125	12.384
5	0.398	0.125	4.132	5	0.398	0.125	9.881
6	0.398	0.125	5.092	6	0.398	0.125	9.107
7	0.607	0.180	4.317	7	0.607	0.180	7.379

Table E.14.1 (continued)

	T = 120 °C				T = 135 °C		
N^o	p_{H_2} (atm)	p_T (atm)	$10^6 \cdot R$ (mol/(kg·s))	N^o	p_{H_2} (atm)	p_T (atm)	$10^6 \cdot R$ (mol/(kg·s))
8	0.607	0.180	4.393	8	0.607	0.180	6.695
9	0.605	0.214	3.029	9	0.605	0.214	5.676
10	0.605	0.214	3.318	10	0.605	0.214	4.747
11	0.600	0.266	2.232	11	0.600	0.266	3.278
12	0.600	0.266	1.714	12	0.600	0.266	3.884
13	0.609	0.323	1.560	13	0.609	0.323	2.505
14	0.609	0.323	1.560	14	0.609	0.323	2.376

*Exercise inspired by the work of Lindfors et al., *Ind. Eng. Chem. Res.*, 32, 34–42, 1993.

E.15 Three-phase reactions in a semi-batch reactor

Hydrogenation of a sugar molecule A is conducted in a three-phase slurry reactor (autoclave) over a metal catalyst. The (simplified) overall reaction scheme is

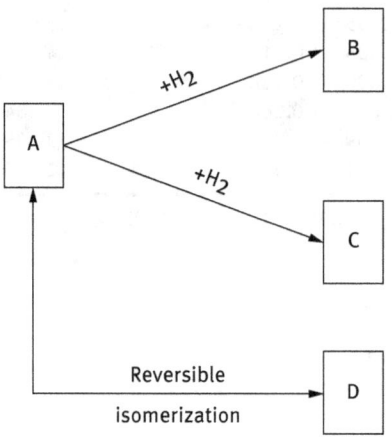

Under the reaction conditions, two hydrogenation products are possible, B and C. Parallel to the hydrogenation, A can undergo isomerization to D.

Assume that the following surface reaction model is valid and leads to the rate equations,

Step 1	$A + * \rightleftharpoons A*$
Step 2	$H_2 + 2* \rightleftharpoons 2H*$
Step 3	$A* + H* \rightarrow B'* + *$
Step 4	$B'* + H* \rightarrow B* + *$
Step 5	$A* + H* \rightarrow C'* + *$
Step 6	$C'* + H* \rightarrow C* + *$
Step 7	$A* \rightleftharpoons D*$
Step 8	$B* \rightleftharpoons B + *$
Step 9	$C* \rightleftharpoons C + *$
Step 10	$D* \rightleftharpoons D + *$

$$R_{B,RDS} = \frac{k_3 K_A \sqrt{K_H} c_A \sqrt{c_{H_2}}}{\left(1 + K_A c_A + \sqrt{K_H c_{H_2}} + K_B c_B + K_C c_C + K_D c_D\right)^2} \quad \text{(E.15.1)}$$

$$R_{C,RDS} = \frac{k_5 K_A \sqrt{K_H} c_A \sqrt{c_{H_2}}}{\left(1 + K_A c_A + \sqrt{K_H c_{H_2}} + K_B c_B + K_C c_C + K_D c_D\right)^2} \quad \text{(E.15.2)}$$

$$R_{D,RDS} = \frac{k_7\left(K_A c_A - \frac{K_D c_D}{K_7}\right)}{1 + K_A c_A + \sqrt{K_H c_{H_2}} + K_B c_B + K_C c_C + K_D c_D} \quad (E.15.3)$$

where * represents an adsorption site, $R_{B,RDS}$, $R_{C,RDS}$ and $R_{D,RDS}$ are the rate of formation of B, C and D respectively.

a) Derive $R_{B,RDS}$, $R_{C,RDS}$ and $R_{D,RDS}$ based on the surface reaction model, assuming that steps 3, 5 and 7 are the rate determining steps of the hydrogenation and isomerization reactions.
b) Assume that the products B, C and D have a very weak affinity to the catalyst surface, and therefore, desorb immediately as they are produced. Rewrite the rate equations (E.15.1–E.15.3) for this particular case.
c) Estimate the parameters of the rate equations derived in b) from the data provided in Table E.15.1. Assume that the reactor is saturated with hydrogen at all times and that no internal or external diffusion limitations exist.

Table E.15.1: Concentration of reactants and products. Experiments performed at 120 °C.

	20 bar			
Time (min)	A (mol/L)	B (mol/L)	C (mol/L)	D (mol/L)
0	1.5	0	0	0
60	1.183	0.2616	0.0447	0.0153
120	0.9539	0.4831	0.0823	0.0299
180	0.5726	0.722	0.121	0.0398
240	0.4333	0.8717	0.1743	0.0462
300	0.2792	0.9339	0.177	0.057
360	0.1645	0.9807	0.1841	0.0527
	40 bar			
Time (min)	A (mol/L)	B (mol/L)	C (mol/L)	D (mol/L)
0	1.5	0	0	0
60	1.1752	0.3036	0.0503	0.015
120	0.7931	0.5172	0.1124	0.0246
180	0.5929	0.8369	0.1479	0.0349
240	0.3429	0.8695	0.17	0.0398
300	0.196	1.1196	0.1845	0.0502
360	0.1167	1.1411	0.2217	0.0463
	60 bar			
Time (min)	A (mol/L)	B (mol/L)	C (mol/L)	D (mol/L)
0	1.5	0	0	0
60	1.1683	0.3321	0.0586	0.0146

Table E.15.1 (continued)

	60 bar			
Time (min)	A (mol/L)	B (mol/L)	C (mol/L)	D (mol/L)
120	0.8623	0.5328	0.099	0.0239
180	0.5604	0.791	0.1496	0.0324
240	0.3199	1.0129	0.1893	0.036
300	0.2037	1.1684	0.197	0.0398
360	0.1221	1.2208	0.1891	0.0437
	80 bar			
Time (min)	A (mol/L)	B (mol/L)	C (mol/L)	D (mol/L)
0	1.5	0	0	0
60	1.1042	0.3144	0.0601	0.0115
120	0.7728	0.6558	0.1116	0.0233
180	0.509	0.8089	0.1338	0.0329
240	0.3497	0.8958	0.1716	0.0314
300	0.2084	1.1102	0.2018	0.0392
360	0.11	1.1795	0.1897	0.0397

Table E.15.2: Data.

Temperature	120 °C
Henry constant(Hydrogen @120°C)	$1.2 \cdot 10^{-5}$ mol/(L·bar)
ρ_B (catalyst loading)	8 g/L

E.16 Non-isothermal liquid phase reaction in a CSTR

The kinetics of the homogeneous liquid-phase decomposition of hydrogen peroxide in presence of thiosulfate

$$2S_2O_3^{2-} + 4H_2O_2 \rightarrow S_3O_6^{2-} + SO_4^{2-} + 4H_2O \tag{E.16.1}$$

was studied in an adiabatic CSTR, operating at transient conditions, by measuring the reactor temperature during the reaction. The temperature data are shown in Table E.16.1. The inlet concentrations of thiosulphate and hydrogen peroxide were determined by chemical analysis prior to the experiments; no chemical analysis was performed during the experiments. At the initial state, the reactor was filled with pure solvent.

Table E.16.1: The experimental data from the decomposition of H_2O_2.

	Experimental data	
$R = 8.3143\,J\cdot K^{-1}\cdot mol^{-1}$		
$c_P = 4.186\,kJ\cdot kg^{-1}\cdot K^{-1}$	t/min	T/°C
$\rho = 1000\,kg\cdot m^{-3}$		
$\Delta H_R = -1004.3\,kJ\cdot mol^{-1}$	0	25.4
$T_0 = 25.2°C$	0.5	27.9
$c_{0A} = 316.8\,mol\cdot m^{-3}, A = S_2O_3^{\,-2}$	1.0	32.9
$c_{0B} = 2c_{0A},\ B = H_2O_2$	1.5	39.6
$V_R = 110\cdot 10^{-6}\,m^3$	2.0	46.1
$V_0 = 130\cdot 10^{-6}\,m^3/min$	2.5	50.4
	3.0	53.1
	4.0	55.5
	4.5	56.0
	5.0	56.2
	5.5	56.5
	6.0	56.3
	6.5	56.5
	7.0	56.5
	7.5	56.4

Reaction (E.16.1) follows approximately a second order kinetics:

$$R = kc_{S_2O_3}c_{H_2O_2} \tag{E.16.2}$$

The heat capacity of the solution is virtually constant during the reaction.
a) In which way the mass and energy balances can be coupled to obtain the reactor temperature as a single variable in the system?
b) Estimate the frequency factor (A) and the activation energy parameter (E_a/R) of the second order rate constant k.

$$k = A \cdot exp\left(-\frac{E_a}{RT}\right) \qquad (E.16.3)$$

by non-linear regression analysis using the following transient temperature data.

E.17 Oxidation of sulphur dioxide in an optimal multi-bed reactor system

The industrial production of sulphuric acid is based on the oxidation of sulphur dioxide on a V_5O_5 –catalyst:

$$SO_2 + \frac{1}{2}O_2 \leftrightarrow SO_3 \qquad (E.17.1)$$

Reaction (E.17.1) is carried out in an adiabatic multibed reactor system which consists of fixed beds with intermediate heat exchangers. The reactor system operates at atmospheric pressure and nitrogen is used as inert gas.

The catalyst is rather inactive at lower temperatures, thus the minimum temperature in the reactor system is 703 K. The conversion of SO_2 at the multibed outlet should be 0.95.

Optimize the intermediate conversions of SO_2 in the reactor system in such a way that the total space time (τ) becomes as short as possible ($\tau = V_R/\dot{V}_0$). The reactor system should consist of three adiabatic beds in series with intermediate heat exchangers.

Table E.17.1: Data.

Reaction kinetics	$R = k_1 p_{SO_2} p_{O_2}^{1/2} - k_2 p_{SO_3}$ $A_1 = 5.412 \text{ mol}/(s \cdot kg \cdot Pa^{3/2})$ $Ea_1 = 129.8 \text{ kJ/mol}$ $A_2 = 7.490 \cdot 10^7 \text{ mol}/(s \cdot kg \cdot Pa)$ $Ea_2 = 224.4 \text{ kJ/mol}$
Reaction thermodynamics	$\Delta H_R = -102.99 \cdot 10^3 + 8.33 \cdot T \text{ [J/mol]}$ $c_p = 1046.7 \text{ J} \cdot kg^{-1} \cdot K^{-1}$
Inlet gas conditions	$T_{01} = 703 \text{ K}$ $P_0 = 101.3 \cdot 10^3$ $p_{ay_{0,SO_2}} = 0.085$ $y_{0,O_2} = 0.090$ $y_{0,N_2} = 0.825$ $\dot{n} = 20.67 \text{ mol/s}$ $M = 31.42 \text{ g/mol}$
Inlet temperatures of second and third beds	$T_{02} = 703 \text{ K}$ $T_{03} = 733 \text{ K}$
Catalyst bulk density	$\rho_B = 600 \text{ kg} \cdot m^{-3}$

E.18 Modelling of a monolith channel

The basic part of an automotive exhaust catalyst, a monolith, consists of a cylindrical channel where the exhaust gas flows and reacts in a thin catalyst layer (see Figure E.18.1).

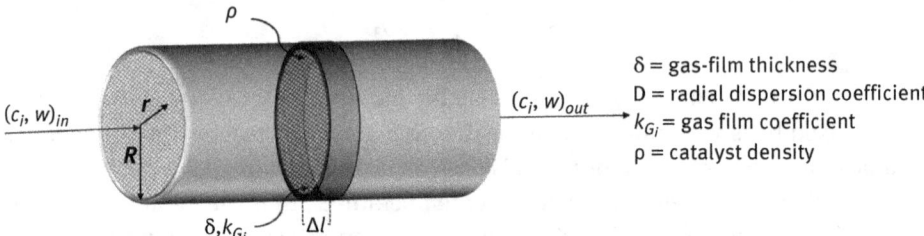

Figure E.18.1: A channel in an exhaust catalyst monolith.

At low flow velocities, laminar conditions prevail and the velocity profile is given by eq. (E.18.1):

$$w(r) = w_0 \left[1 - \left(\frac{r}{R}\right)^2\right] \qquad (E.18.1)$$

where w_0 is the flow velocity at the axis of the channel ($r = 0$). Radial concentration gradients appear spontaneously in the channel due to chemical reactions in the catalyst layer. Because of this, molecules start to diffuse in the radial direction. The gas-film around the catalyst layer retards the mass transport while the catalyst layer itself is so thin that no diffusion limitations appear there. The reaction rate can thus be approximated to be constant inside the catalyst layer.

a) Derive the mass balance for an arbitrary component i in gas-phase at steady state. Use preferably the notation given in Figure E.18.1.
b) Give the boundary conditions for the mass balance.
c) Describe qualitatively how you would solve the mass balance numerically.
d) Develop an extended model, in which the mass transfer limitations in the porous catalyst layer are included.
e) How would you solve d) numerically?

E.19 Heterogeneous two-dimensional model for a catalytic fixed-bed reactor

Derive the mass and energy balances for a catalytic fixed-bed reactor, where both internal and external mass transfer limitations exist in the catalyst particles and radial dispersion in the reactor tube appear. Suggest a suitable numerical strategy for solving this problem.

E.20 Dissolution of a solid particle in a batch reactor

Assume that a solid component (A) dissolves in a liquid where it reacts with a liquid phase component (B) to a dissolved product (P). Stirring in the reactor is not particularly efficient, which leads to the formation of a film around the particle. The chemical reaction proceeds both in the liquid film and in the liquid bulk. Derive the mass balance equations of the components in the film and in the liquid bulk. Take in to account the fact that the film thickness changes as the reaction progresses because the size of the solid particle diminishes.

Appendices

A.1 Numerical strategies in the solution of non-linear algebraic equations and ordinary differential equations

Two kinds of numerical tasks appear very often in solving tasks in chemical reaction engineering: non-linear algebraic equations and ordinary differential equations. For instance, the steady-state continuous stirred tank reactor (CSTR) model for a multicomponent and multireaction system is in general a set of non-linear equations (NLEs); only in case of an isothermal reactor and first-order kinetics a set of linear equations is obtained. The steady-state plug flow reactor model is described by ordinary differential equations (ODEs). The inlet conditions are known, and the problem can be solved numerically forward, having the reactor length, the reactor volume or the space time or the residence time as the independent variable. The dynamic CSTR model, which describes the transient behaviour of the reactor, is from the numerical viewpoint a straightforward ODE model, an initial value problem (IVP): the initial concentrations and temperature in the tank are known and the ODEs are solved numerically as a function of the real time. Many other models appearing in chemical reaction engineering can be transformed to ODEs. Typical examples are transient models for plug flow reactors as well as tubular reactors described with the axial dispersion model. The spatial coordinate (length coordinate) can be discretized by finite differences or by applying finite element methods (e.g. orthogonal collocation) and the PDEs are thus converted to a large set of ODEs, again an initial value problem. Some numerical algorithms for ODEs and NLEs are described in detail in the textbook of Salmi et al. (2019). Here, a brief summary of the methodology is given.

A.1.1 Non-linear algebraic equations

We consider a non-linear algebraic equation system of unknown variables (y)

$$f(x, y) = 0 \qquad (A.1.1)$$

In eq. (A.1.1), x denotes a continuity variable (e.g. the space time, $\tau = V_R/\dot{V}$).

A powerful approach for the numerical solution of the system (1) is the multidimensional Newton-Raphson method, which has good convergence properties, provided that the initial guess of the solution is close enough to the solution. The multidimensional Newton-Raphson algorithm is written as

$$y_k = y_{k-1} - \underline{J}^{-1} f_{k-1} \qquad (A.1.2)$$

where k denotes the iteration index, f_k, is the function vector and $\underline{J}^{-1}$ is the inverted Jacobian matrix containing the partial derivatives of the functions f_i.

An efficient way to carry out the inversion of J is Gaussian elimination. The derivatives $\partial f_i/\partial y_j$ are obtained either from analytical expressions or by numerical approximation. The approximation of $\partial f_i/\partial y_j$ can be done with forward differences accordingly,

$$\frac{\partial f_i}{\partial y_j} \approx \frac{\Delta f_i}{\Delta y_j} = \frac{f_i(y_j + \Delta y_j) - f_i(y_j)}{\Delta y_j} \qquad (A.1.3)$$

The choice of difference Δy_j is critical: the smaller the Δy_j is, the more accurate the approximation of the derivative, but if the accuracy of the computer is exceeded, the denominator of eq. (A.1.3) becomes zero. A criterion for the selection of Δy_j, is

$$\Delta y_j = max(\varepsilon_c, \varepsilon_R|y_j|) \qquad (A.1.4)$$

where ε_R is the relative difference given by the program user and ε_c is the round-off criterion of the computer.

The initial estimate, y_0, should be carefully chosen for a successful iteration according to the Newton-Raphson algorithm. In many cases, the problem is formulated in such a way that the solution is desired as a function of the continuity parameter (x), the mass balances of a CSTR should be solved as a function of the space time $(\tau = V_R/\dot{V})$. In such cases, x can be selected as the continuity parameter: the solution of the equation system (1) obtained for one parameter value of $x(y_0)$ can be used as an initial estimate for the solution of the problem for the subsequent parameter value $x + \Delta x$:

$$y_0(x + \Delta x) = y_\infty(x) \qquad (A.1.5)$$

In this way, one can proceed with respect to the continuity parameter, x. The convergence of the Newton-Raphson algorithm is usually quadratic, which guarantees rather rapid computations. The Newton-Raphson iteration is stopped as a sufficient accuracy is achieved, typically when the changes of the successive iterations approach zero.

A.1.2 Ordinary differential equations

For first-order ordinary differential equations

$$\frac{dy}{dx} = f(y, x) \qquad (A.1.6)$$

many numerical methods have been solved since the pioneering work of the great German mathematician Leonard Euler. The following initial condition is assumed,

$$y = y_0 \text{ vid } x = x_0 \qquad (A.1.7)$$

In classical literature of numerical methods devoted to the solution of ODEs, initial value problems Euler's method and explicit Runge-Kutta methods are treated in detail. The drawback of these explicit methods is that they cannot be successfully used as as a general approach to simulation tasks appearing in chemical reaction engineering. A very characteristic feature for problems in reaction engineering is stiffness, which is typically originated from very different values of the rate constants appearing in the system: some of the reactions can be very rapid, close to equilibria, while other ones can be extremely slow. An ultimate example of such cases is a gas-phase system incorporating radical reactions. Non-isothermal conditions and discretization procedures applied to PDEs can increase the stiffness of the system.

Mathematicians have developed robust algorithms for stiff differential equations, such as semi-implicit Runge-Kutta methods and linear multistep methods (1–7). A brief summary of some algorithms is given below.

A.1.2.1 Semi-implicit Runge-Kutta methods

Runge-Kutta methods have the general form (1),

$$y_n = y_{n-1} + \sum_{i=1}^{q} b_i k_i \tag{A.1.8}$$

where y_n and y_{n-1} give the solution of the differential eq. (A.1.1) at x and $x + \Delta x$. The coefficient vector is obtained from

$$k_i = h \cdot f\left(x_n, y_n + \sum_{l=1}^{i} a_{il} k_l\right) \tag{A.1.9}$$

where $h = \Delta x$.

If the coefficient $a_{il} = 0$ for $l = i$ the method is explicit; if $a_{il} \neq 0$ for $l = i$ the method is implicit.

$$(I - Ja_{ii}h)k_i = h \cdot f\left(y_n + \sum_{l=1}^{i-1} a_{il} k_l\right) \tag{A.1.10}$$

where I denotes the identity matrix:

$$I = \begin{bmatrix} 1 & 0 & \cdots & 0 \\ 0 & 1 & \cdots & 0 \\ \vdots & \vdots & \ddots & \vdots \\ 0 & 0 & 0 & 1 \end{bmatrix} \tag{A.1.11}$$

and J is the Jacobian matrix:

$$J = \begin{bmatrix} \partial f_1/\partial y_1 & \partial f_1/\partial y_2 & \cdots & \partial f_1/\partial y_N \\ \partial f_2/\partial y_1 & \partial f_2/\partial y_2 & \cdots & \partial f_2/\partial y_N \\ \vdots & \vdots & \ddots & \vdots \\ \partial f_N/\partial y_1 & \partial f_2/\partial y_2 & \cdots & \partial f_{np}/\partial y_N \end{bmatrix} \qquad (A.1.12)$$

Analytical expressions for the partial derivatives in the Jacobian can be used, or alternatively, they are estimated by using finite differences

$$\frac{\partial f_i}{\partial y_j} = \frac{f_i(y_j + \Delta y_j) - f_i(y_j)}{\Delta y_j} \qquad (A.1.13)$$

After developing of $\underline{k_i}$ in a Taylor series around $y_n + \sum_{l=1}^{i-1} a_{il} k_l$ and truncating after the first term a semi-implicit Runge-Kutta method is obtained, according to which $\underline{k_i}$ is calculated from

$$(I - J a_{ii} h) k_i = h \Delta f \left(y_n + \sum_{l=1}^{i-1} a_{il} k_l \right) + \sum_{l=1}^{i-1} c_{il} k_l \qquad (A.1.14)$$

An extension of semi-implicit Runge-Kutta method is the Rosenbrock-Wanner method (ROW) (4). Coefficients for a fourth-order Rosenbrock-Wanner method are listed in Table A.1.1.

Table A.1.1: Coefficients for the ROW4A-method.

$a_{ii} = 0.395$	
$a_{21} = 0.438$	$c_{21} = -1.943744189$
$a_{31} = 0.9389486785$	$c_{31} = 0.4169575310$
$a_{32} = 0.0730795421$	$c_{32} = 1.3239678207$
$b_1 = 0.72904488$	$c_{41} = 1.5195132578$
$b_2 = 0.0541069773$	$c_{42} = 1.3537081503$
$b_3 = 0.2815993624$	$c_{43} = -0.8541514953$
$b_4 = 0.25$	

Kaps and Wanner (1981).

The linear equation system (10) is solved every time in the calculations of $\underline{k_i}$. The matrix $I - J a_{ii} h$ is inverted by Gaussian elimination. After obtaining $\underline{k_i}$ according to eq. (A.1.5), y_n is easily calculated from (3). The coefficients a_{il} and $\underline{k_i}$ for some other semi-implicit Runge-Kutta methods, such as the methods of Rosenbrock, Caillaud-Padmanabhan and Michelsen are listed in the original literature (1–3) and in the Salmi et al. (2019).

A.1.2.2 Linear multistep methods

Two kinds of multistep methods are used: Adams-Moulton and backward difference methods. The Adams-Moulton originates back to old times, whereas the first systematic approach to backward difference methods was presented by Henrici in 1962. For linear multistep methods, the following algorithm is valid for the solution of the ordinary differential equations presented by eq. (A.1.1):

$$y_n = \sum_{i=1}^{K_1} \alpha_i y_{n-1} + h \cdot \sum_{i=0}^{K_2} \beta_i f_{n-i} \qquad (A.1.15)$$

Depending on the values of K_1 and K_2, Adams-Moulton (AM) and backward-difference methods (BD) are obtained (5–6):

$$\alpha_1 = 1 \ K_1 = 1 \ K_2 = q - 1 \ (AM) \qquad (A.1.16)$$

$$K_1 = q \ K_2 = 0 \ (BD) \qquad (A.1.17)$$

Application of the conditions (16) and (17) on eq. (A.1.15) gives the Adams-Moulton method,

$$y_n = y_{n-1} + h \cdot \sum_{i=0}^{q-1} \alpha_i f_{n-i} \ (AM) \qquad (A.1.18)$$

and the backward-difference method:

$$y_n = \sum_{i=1}^{q} \alpha_i y_{n-i} + h \beta_0 f_{n-i} \ (BD) \qquad (A.1.19)$$

Both methods are implicit by their nature f_n usually is a non-linear function of y_n. Equations (A.1.18) and (A.1.19) imply a solution of a non-linear algebraic equation system of the type

$$g_n = y_n - h \beta_0 f_n - a_n = 0 \qquad (A.1.20)$$

where a_n is given by

$$a_n = y_{n-1} + h \cdot \sum_{i=1}^{q-1} \beta_i f_n \ (AM) \qquad (A.1.21)$$

or by

$$a_n = \sum_{i=1}^{q} \alpha_i y_{n-i} \ (BD) \qquad (A.1.22)$$

For the solution of y_n, a_n is known from the previous solutions; the problem consists thus of an iterative solution of the equation system (20). This can be done by using the Newton-Raphson method,

$$y_{n(k+1)} = y_{n(k)} - P_n^{-1} g_n \qquad (A.1.23)$$

where P_n is the Jacobian. Differentiation of eq. (A.1.23) gives a useful expression for P_n

$$P_n = I - h\beta_0 J \qquad (A.1.24)$$

where the identity matrix I and the Jacobian are given by eqs. (A.1.11) and (A.1.12), respectively. The coefficients α_i and β_i for Adams-Moulton and backward-difference networks of various orders are listed in Tables A.1.2 and A.1.3. The simplest backward difference method, the two-point formula, is called implicit Euler's method. For stiff differential equations, the backward difference algorithm should be preferred to the Adams-Moulton method (5–7).

Table A.1.2: β-coefficients in the Adams-Moulton method.

i	0	1	2	3	4	5
β_{1i}	1					
$2\beta_{2i}$	1	1				
$12\beta_{3i}$	5	8	−1			
$24\beta_{4i}$	9	19	−5	1		
$720\beta_{5i}$	251	646	−264	106	−19	
$1440\beta_{6i}$	475	1427	−798	482	173	27

Table A.1.3: α-coefficients in the backward difference method.

i	0	1	2	3	4	5	6
α_{1i}	1	1					
$3\alpha_{2i}$	2	4	−1				
$11\alpha_{3i}$	6	18	−9	2			
$25\alpha_{4i}$	12	48	−36	16	−3		
$137\alpha_{5i}$	60	300	−300	200	−75	12	
$147\alpha_{6i}$	60	360	−450	400	−225	72	−10

$\alpha_{10} = \beta_{10}$, $\alpha_{20} = \beta_{20}$, etc.

For the solution of stiff differential equations with linear multistep methods, the well-known code LSODE with different options was published in 1980s by A. Hindmarsh (1983). The code is very efficient, and later on, different variations of it have been developed, for instance, a version for sparse systems (LSODEs).

References

Caillaud, J. B., & Padmanabhan, L. (1971). Chemical Engineering Journal, 2, 227.
Gear, C. W. (1971). Numerical Initial Value Problems in Ordinary Differential Equations, Englewood Cliffs, N. J.: Prentice Hall.
Henrici, P. (1962). Discrete Variable Methods in Ordinary Differential Equations, New-York: Wiley.
Hindmarsh, A. C. (1983). ODEPACK, A Systematized Collection of ODE-Solvers in Scientific Computing (R. Stepleman, et al., Eds.), 55–64, IMAC/North Holland Publishing Company.
Kaps, P., & Wanner, G. (1981). Numerische Mathematik, 38, 279.
Michelsen, M. L. (1976). AIChE. Journal, 22, 594.
Rosenbrock, H. H. (1963). The Computer Journal, 5, 329.
Salmi, T., Mikkola, J.-P., & Wärnå, J. (2019). Chemical Reaction Engineering and Reactor Technology, Boca Raton, Florida: Taylor & Francis CRC Press, (2nd edition).

A.2 Computer simulation of CSTR, PFR and batch reactor models

Example 1

CSTR Batch reactor Solve the algebraic equation system,

$$y_{01} - y_1 - r_1 = 0$$
$$y_{02} - y_2 - r_2 = 0$$
$$y_{03} - y_3 + r_1 - r_2 = 0$$
$$y_{04} - y_4 + r_2 = 0$$
$$y_{05} - y_5 + r_1 + r_2 = 0$$

which describes a consecutive-competitive reaction, A+B→R+E, R+B→S+E, in a CSTR. The reaction kinetics is defined by

$$r_1 = k_1 y_1 y_2 x$$
$$r_2 = k_2 y_3 y_2 x$$

where x denotes the residence time. The parameters $y_{01}, y_{02}, \ldots, y_{05}$ as well as k_1 and k_2 have constant values. The value of the parameter x (residence time) should reside in between 0 and 5. A Matlab code for solving this problem is listed below and the concentration profiles of components 1(=A), 2(=B), 3(=R), 4(=S) and 5(=E) are displayed in Figure A.2.1 and the Python and Matlab codes are listed as follows

```python
# Appendix A2 example 1
# A+B->R+E
# R+B->S+E
# Python version 3.7, Spyder version 3.3.3
import numpy as np
import matplotlib.pyplot as plt
from scipy.optimize import fsolve

y0 = [1, 1, 0, 0, 0]    # Initial values A and B =1 R, S and E zero
ys =np.zeros((100,5)) # result vector
k1=5   # rate constants for reaction 1 and 2
k2=2
```

```python
# Define a function with the CSTR mass balance
def func(state,tau):
    cA=state[0] # concentrations in reactor
    cB=state[1]
    cR=state[2]
    cS=state[3]
    cE=state[4]
    c0A=y0[0] # inlet concentrations
    c0B=y0[1]
    c0R=y0[2]
    c0S=y0[3]
    c0E=y0[4]
    r1=k1*cA*cB*tau # reaction rates
    r2=k2*cR*cB*tau
    zeroA= (c0A-cA)-r1    # A  CSTR mass balances for each component
    zeroB= (c0B-cB)-r1-r2 # B
    zeroR= (c0R-cR)+ r1-r2 # R
    zeroS= (c0S-cS)+ r2    # S
    zeroE= (c0E-cE)+r1+r2  # E
    return [zeroA, zeroB, zeroR, zeroS, zeroE]

tau = np.linspace(0,5,100)  # residence time from 0 to 5 with 100 points
for i in range(0,100,1):
    y = fsolve(func, y0,tau[i])
    ys[i,:]=y[:]

plt.figure(1)
plt.plot(tau, ys[:,0],'b-',label='A')
plt.plot(tau, ys[:,1],'r-',label='B')
plt.plot(tau, ys[:,2],'g-',label='R')
plt.plot(tau, ys[:,3],'m-',label='S')
plt.plot(tau, ys[:,4],'k-',label='E')
plt.legend(loc='upper right', shadow=True, fontsize='large')
plt.xlabel(r'$\tau$')
plt.ylabel('concentration')
plt.grid()
```

Matlab

```
% Appendix A2 example 1
% A+B->R+E
% R+B->S+E
```

```matlab
function A21
global y0 k1 k2 tau
y0 = [1, 1, 0, 0, 0];   % Initial values A and B =1 R,S and E zero
k1=5;   % rate constants for reaction 1 and 2
k2=2;
x=(0:.05:5)
for i=1:101
  tau=x(i);
y(i,:)=fsolve(@cstr,[1 1 0 0 0]);
end
plot(x,y);
xlabel('residence time');ylabel('concentration');
grid on;
legend('A','B','R','S','E');
end

function zero=cstr(y)
global y0 k1 k2 tau
cA=y(1); %concentrations in reactor
cB=y(2);
cR=y(3);
cS=y(4);
cE=y(5);
c0A=y0(1); % inlet concentrations
c0B=y0(2);
c0R=y0(3);
c0S=y0(4);
c0E=y0(5);
r1=k1*cA*cB*tau; %reaction rates
r2=k2*cR*cB*tau;
zero(1,1)= (c0A-cA)-r1;      % A   CSTR mass balances for each component
zero(2,1)= (c0B-cB)-r1-r2;   % B
zero(3,1)= (c0R-cR)+r1-r2;   % R
zero(4,1)= (c0S-cS)+r2;      % S
zero(5,1)= (c0E-cE)+r1+r2;   % E
end
```

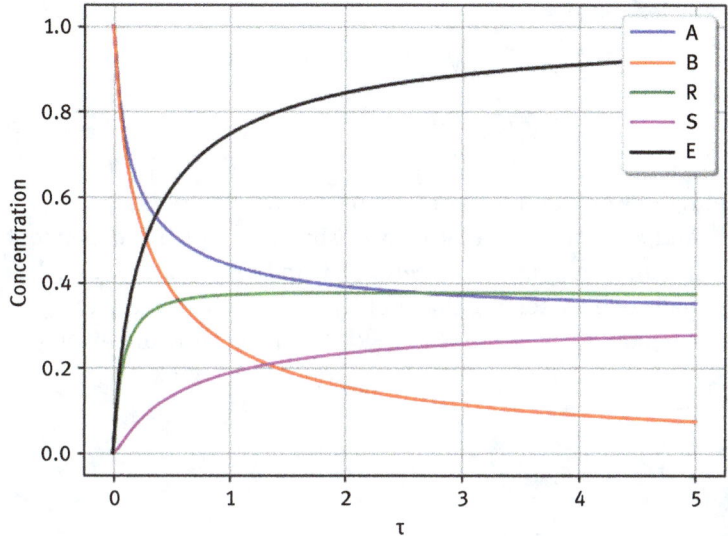

Figure A.2.1: Simulated concentration profiles of Example 1, steady-state CSTR, τ=residence time.

Example 2

Solve the following differential equations,

$$\frac{dy_1}{dx} = -r_1$$

$$\frac{dy_2}{dx} = -r_1 - r_2$$

$$\frac{dy_3}{dx} = r_1 - r_2$$

$$\frac{dy_4}{dx} = r_2$$

$$\frac{dy_5}{dx} = r_1 + r_2$$

with the initial conditions at $x = 0$,

$$y_{01} = 1.0$$

$$y_{02} = 1.0$$

$$y_{03} = y_{04} = y_{05} = 0$$

The terms, r_1 and r_2 are defined as

$$r_1 = k_1 y_1 y_2$$

$$r_2 = k_1 y_3 y_2$$

The equations describe a consecutive-competitive reaction, $A + B \rightarrow R + E$, $R + B \rightarrow S + E$, in a plug flow or batch reactor (compare with Example 1).

The parameters k_1 and k_2 obtain constant values. The value of the independent variable, x (residence time), should be varied between 0 and 5 (in arbitrary units).

A Python and a Matlab code for solving this problem is listed below and the concentration profiles of components 1(=A), 2(=B), 3(=R), 4(=S) and 5(=E) are displayed in Figure A.2.2.

Python
```
# Appendix A2 example 2
# A+B->R+E
# R+B->S+E
# Python version 3.7, Spyder version 3.3.3
import numpy as np
import matplotlib.pyplot as plt
from scipy.integrate import odeint
y0 = [1, 1, 0, 0, 0]  # Initial values A and B =1 R,S and E zero
k1=5  # rate constants for reaction 1 and 2
k2=2
# Define a function which calculates the derivative
def dcdt(state, t):
    cA=state[0]
    cB=state[1]
    cR=state[2]
# reaction rates
    r1=k1*cA*cB
    r2=k2*cR*cB
# mass balances for each component in a batch reactor
    dcAdt=r1         # A
    dcBdt=r1-r2      # B
    dcRdt= r1-r2     # R
    dcSdt= r2        # S
    dcEdt= r1+r2     # E
    return [dcAdt, dcBdt, dcRdt, dcSdt, dcEdt]

xs = np.linspace(0,5,100)
ys = odeint(dcdt, y0, xs)
```

```python
plt.figure(1)
plt.plot(xs, ys[:,0],'b-',label='A')
plt.plot(xs, ys[:,1],'r-',label='B')
plt.plot(xs, ys[:,2],'g-',label='R')
plt.plot(xs, ys[:,3],'m-',label='S')
plt.plot(xs, ys[:,4],'k-',label='E')
plt.legend(loc='upper right', shadow=True, fontsize='large')
plt.xlabel('residence time')
plt.ylabel('concentration')
plt.grid()
```

Matlab

```matlab
% A+B->R+E
% R+B->S+E
function A22
% solve ode from time=0 to time=5 with
% initial concentrations A=1, B=1 and R,S,E=0
[x,y]=ode45(@pfr,[0 5],[1 1 0 0 0]);
plot(x,y);
xlabel('residence time');ylabel('concentration');
grid on;
legend('A','B','R','S','E');
end

function dy=pfr(t,y)
cA=y(1);
cB=y(2);
cR=y(3);
k1=5;
k2=2;
r1=k1*cA*cB;
r2=k2*cR*cB;
dy(1,1)= -r1;        % A
dy(2,1)= -r1-r2;     % B
dy(3,1)=  r1-r2;     % R
dy(4,1)=  r2;        % S
dy(5,1)= r1+r2;      % E
end
```

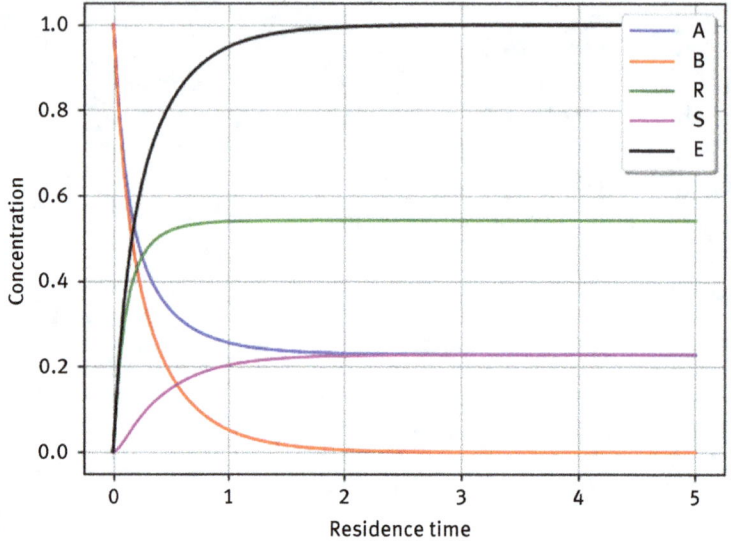

Figure A.2.2: Simulated concentration profiles of Example 2, steady-state PFR.

A.3 Numerical simulation of non-isothermal tubular reactors

A dimerization reaction in gas phase

$$2A \rightarrow P$$

is carried out in a tubular reactor with plug flow. The molar flows and the reactor temperature should be simulated as a function of the reactor volume. The reaction is elementary and the data relevant for the simulation are listed below.

Data

$$\dot{n}_{0A} = 0.1 \; mol/s$$

$$\dot{n}_{0P} = 0 \; mol/s$$

$$T_0 = 473 \; K$$

$$\dot{m} = 2.4 \cdot 10^{-3} \; Kg/s$$

$$c_P = 1.5 \cdot 10^3 \; J/Kg \cdot K$$

$$\Delta H_r = -50.0 \cdot 10^3 \; J/mol$$

$$UA/V_r = 45 \; J/m^3 \cdot s \cdot K$$

$$x_{0A} = 1.0$$

$$\delta_A = \frac{(\nu_A + \nu_P)}{-\nu_A} = \frac{(-2+1)}{2} = -\frac{1}{2}$$

$$A = e^{1.5} \; m^3/mol/s$$

$$E_a = 50.0 \cdot 10^3 \; J/mol$$

$$\rho_0 = 1.2 \; Kg/m^3$$

$$\dot{V}_0 = \frac{\dot{m}}{\rho_0}$$

The molar flows are determined by the mass balances

$$\frac{d\dot{n}_A}{dV} = r_A = \nu_A \cdot r \tag{A.3.1}$$

$$\frac{d\dot{n}_P}{dV} = r_P = \nu_P \cdot r \tag{A.3.2}$$

where $\nu_A = -2$ and $\nu_P = 1$.

Since the reaction is irreversible and elementary, the reaction rate is given by

$$r = k \cdot c_A^2 \tag{A.3.3}$$

which can be expressed in the form

$$r = A \cdot e^{-Ea/(RT)} \cdot \left(\frac{\dot{n}_A}{\dot{V}}\right)^2 \tag{A.3.4}$$

The volumetric flow rate is updated with the formula

$$\frac{\dot{V}}{\dot{V}_0} = (1 + x_{0A}\delta_A\eta_A)\frac{T}{T_0} \tag{A.3.5}$$

where

$$\eta_A = \frac{\dot{n}_{0A} - \dot{n}_A}{\dot{n}_{0A}} = 1 - \frac{\dot{n}_A}{\dot{n}_{0A}} \tag{A.3.6}$$

The energy balance for a non-isothermal tubular reaction is given by

$$\frac{dT}{dV} = \frac{1}{\dot{m}c_P}\left(r(-\Delta H_r) - \frac{UA}{V_R}(T - T_C)\right) \tag{A.3.7}$$

The mathematical model consists of eqs. (A.3.1, A.3.2) and (A.3.4–A.3.7). The differential eqs. (A.3.1, A.3.2, A.3.7) are solved numerically together with the initial conditions at

$$\dot{n}_A = \dot{n}_{0A}$$

$$\dot{n}_P = \dot{n}_{0P} = 0$$

$$T = T_0$$

$V = 0$, that is, at the reactor inlet.

A MATLAB and a Python code are shown below that simulates the molar flows and temperature in the reactor

Matlab

```
% A dimerization 2A --> P in a gas phase PFR

clear all   % clear all variable previously used
close all   % close all open windows
clc         % clears workspace screen

VolumeSpan=linspace(0,1.5,100);    % defines the volume domain
```

```matlab
Init0=[0.1 0 473];              % defines the ODE initial conditions
%The molar and the temperature profiles are calculated:
[x,y]=ode45(@ddt, VolumeSpan,Init0);
%The molar flows are plotted with:

hold on
plot(x,y(:,1))
text(1.3,.01,'A')
plot(x,y(:,2))
text(1.3,.05,'P')
xlabel('Volume (m3)')
ylabel('mol/s')
grid on
hold off

%The temperature profile is plotted with:

figure(2);
plot(x,y(:,3))
xlabel('Volume (m3)')
ylabel('T (K)')
grid on

function rhs=ddt(v,n)

% reaction engineering date
n0A=0.1;     % mol/s
m=2.4e-3;    % kg/s
Cp=1.5e3;    % J/kgK
dH=-50.0e3;  % J/mol
T0=473;      % K
Tc=T0;       % K
X0A=1.0;
dA=-0.5;
A=exp(1.5);  %m^3/mol/s
Ea=50.0e3;   % J/mol
rho=1.2;     % kg/m^3
R=8.3143;    % J/mol/K
UAVR=45;     % J/Km^2s

% volumetric flow rate
V0=m/rho;    % m^3/s
```

```matlab
% molar flows and temperature
nA=n(1);
nP=n(2);
T=n(3);

% conversion
etaA=1-nA/n0A;

%update of volumetric flow rate
Vs=(1+X0A*dA*etaA)*T/T0*V0;

%reaction rate
r=A*exp(-Ea/R/T)*(nA/Vs)^2;

%Stoichiometric matrix
SM=[-2 1];

% mass balances
rhs=SM.'*r;
% energy balance

rhs(3,1)=1/m/Cp*(r*(-dH)-UAVR*(T-Tc));
end
```

Python

```python
# Appendix A3
# Python version 3.7, Spyder version 3.3.3
import numpy as np
import matplotlib.pyplot as plt
from scipy.integrate import odeint
from math import exp

# reaction engineering date
n0A=0.1      # inlet flow of A mol/s
n0P=0.0      # inlet flow of P mol/s
m=2.4e-3     # mas flow kg/s
Cp=1.5e3     # Heat capacity J/kgK
dH=-50.0e3   # Heat of reaction J/mol
T0=473       # feed temperature K
Tc=T0        # Coolant temperature K
```

A.3 Numerical simulation of non-isothermal tubular reactors

```python
X0A=1.0      # fraction of A in feed (only A)
dA=-0.5      # Change in molarity due to reaction
A=exp(1.5)   # frequency factor m^3/mol/s
Ea=50.0e3    # Activation energy J/mol
rho=1.2      # Density kg/m^3
R=8.3143     # Gas constant J/mol/K
UAVR=45      # Heat transfer J/Km^2s
VolumeSpan=np.linspace(0,1.5,100) # defines the volume domain
Init0=[n0A, n0P, T0]              # defines the input flow rates and temperature

def ds(state, t) : # Define a function which calculates the derivative

    nA=state[0] # molar flow of A
    nP=state[1] # molar flow of P
    T=state[2]  # temperature
    V0=m/rho                    # feedflow rate m^3/s
    etaA=1-nA/n0A               #conversion
    Vs=(1+X0A*dA*etaA)*T/T0*V0  #update of volumetric flow rate
    r=A*exp(-Ea/R/T)*(nA/Vs)**2 #reaction rate
# mass balances for reactant A and product P
    dnA=-2*r
    dnP= r
# energy balance
    dT=1/m/Cp*(r*(-dH)-UAVR*(T-Tc))
    return [dnA, dnP, dT]
xs = np.linspace(0,5,100)
ys = odeint(ds, Init0, VolumeSpan)

plt.figure(1)
plt.plot(xs, ys[:,0],'b-',label='A')
plt.plot(xs, ys[:,1],'r-',label='P')
plt.grid()
plt.legend(loc='upper right', shadow=True, fontsize='large')

plt.xlabel("Volume")
plt.ylabel("flowrate (mol/s)")
plt.figure(2)
plt.plot(xs, ys[:,2],'r-',label='Temperature')
plt.grid()
plt.xlabel("Volume")
plt.ylabel("Temperature (K)")
```

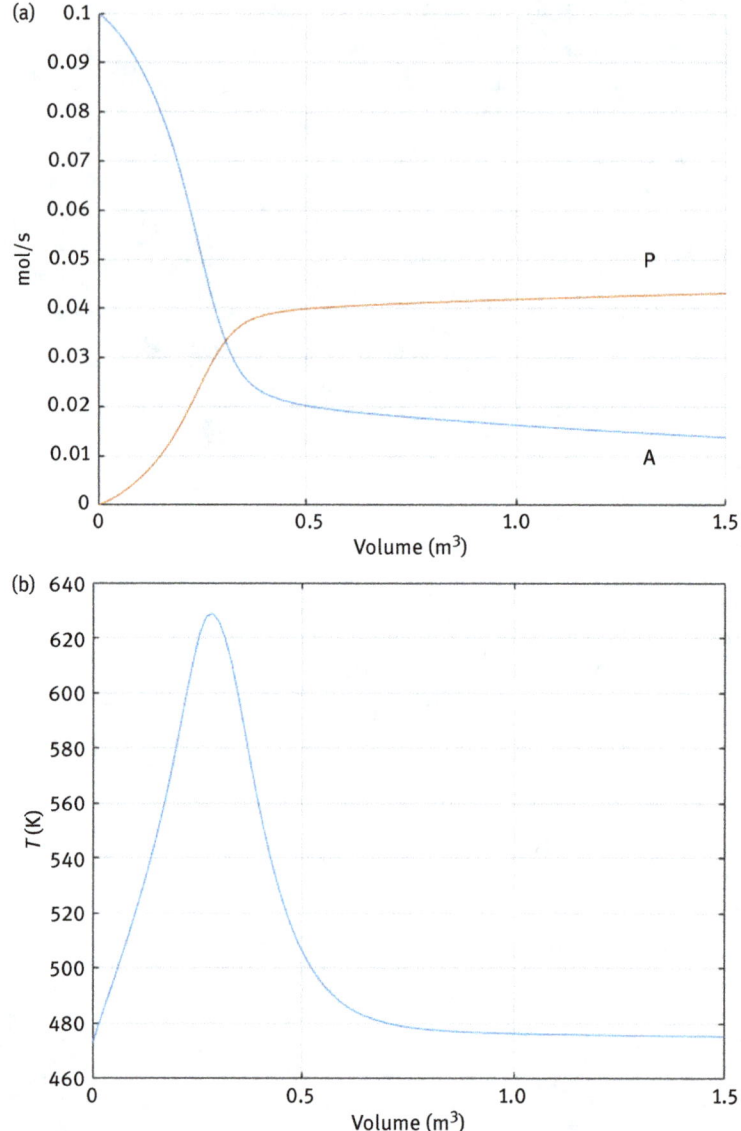

Figure A.3.1: The molar flows and temperature as a function of the reactor volume (in m³) for a tube reactor.

Index

Absorption 1, 97, 98, 118
Absorption column 99
Activity XVI, 7, 118
Adams-Moulton method 215, 216
Adiabatic operation 176, 203
Adiabatic reactor 205
ADM. See axial dispersion model
Adsorption 1, 8, 143, 186, 197, 201
Adsorption Parameter 8, 143, 197
AM. See Adams-Moulton method
Amount of substance XV, 12, 13, 28, 42, 55, 150
Analytical solution V, 18, 31, 58, 91, 92, 105, 126, 128, 131, 180
Arrhenius law 142, 146, 147
Association factor XVI, 115
Axial dispersion 9, 14, 15, 16, 17, 18, 19, 20, 30, 31, 33, 47, 68, 84, 87, 91, 93, 94, 108, 119, 175, 189, 211
Axial dispersion model 9, 15, 16, 17, 18, 19, 20, 30, 31, 68, 84, 91, 94, 211

Backmix reactor. See continuous stirred tank reactor
Backmixing 9, 10, 11, 15, 17, 68, 88, 109, 111, 130
Backward difference 31, 32, 66, 75, 76, 215, 216
Backward difference method 66, 75, 215, 216
Batch reactor 10, 11, 12, 13, 26, 27, 32, 91, 92, 107, 123, 124, 125, 126, 131, 132, 148, 150, 153, 176, 191, 208, 218
BD. See Backward difference method
Berl saddle 99
BET, Brunauer-Emmett-Teller equation 77
Biot number 59
Bodenstein number XVI, 17, 111
Boundary value problem 17, 32, 66, 93, 94, 112, 118
Bubble column 81, 82, 84, 98, 122
Bubble phase 71, 74, 75, 76
Bulk density XVI, 48, 50, 71, 92, 131, 153, 179, 205
Bulk phase 46, 57, 60, 62, 66, 70, 83, 99, 100, 107, 121
Burke-Plummer equation 64
BVP. See boundary value problem

Caillaud-Padmanabhan method 214
Catalysis 41, 143
Catalyst XVI, XVII, 5, 9, 20, 41, 42, 43, 44, 45, 46, 47, 48, 50, 52, 54, 55, 59, 62, 63, 64, 66, 67, 68, 69, 70, 71, 73, 76, 77, 78, 81, 82, 84, 89, 90, 91, 92, 94, 128, 131, 153, 157, 160, 178, 179, 180, 181, 183, 185, 186, 195, 197, 200, 201, 205, 206, 207
Catalyst particle 41, 42, 43, 45, 46, 47, 52, 54, 55, 59, 62, 63, 64, 66, 67, 68, 70, 76, 77, 81, 84, 89, 90, 91, 128, 131, 180, 207
Central difference 32, 33, 65
Chemical industry 41, 45, 68
Cloud phase 70, 71, 72, 74, 75
Collocation 65, 66, 95, 119, 121
Column 85, 93, 98, 99, 107, 108, 109, 113, 118, 119, 121
Column packing 98, 99
Column reactor 85, 93, 107, 108, 109, 113, 118, 119, 121
Complex reaction system V
Composite reactions. See multiple reactions
Compressibility factor 13
Concentration gradient 17, 44, 46, 63, 84, 129, 206
Confidence interval 141
Consecutive-competitive reaction 218
Consumption rate 162
Continuity parameter 35, 212
Continuous stirred tank reactor 9, 11, 177, 187, 211
Contour plot 141, 142, 143
Convergence 35, 36, 37, 95, 139, 211, 212
Conversion 5, 17, 41, 51, 98, 129, 176, 178, 182, 183, 184, 185, 205
Correlation between parameters 141, 142, 143, 145, 147
Covariance 140
CSTR 9, 11, 12, 13, 15, 32, 34, 36, 88, 93, 98, 127, 128, 130, 177, 182, 203, 211, 212, 218, 221
Cylinder 43, 55

DADM. See dynamic axial dispersion model
DAE. See differential-algebraic system
Danckwerts boundary condition 86, 88

DASSL 75
Dependent variable XVI, 12, 13, 49, 132, 133, 134, 135, 144
Desorption 201
Differential method 123, 124, 125, 126
Differential reactor 129, 130, 132, 143, 197
Diffusion XV, 15, 33, 37, 41, 42, 43, 44, 45, 46, 47, 52, 54, 55, 58, 60, 68, 76, 77, 78, 81, 83, 90, 100, 102, 113, 114, 115, 116, 117, 121, 128, 158, 180, 181, 185, 201, 206
Diffusion coefficient 15, 76, 77, 78, 113, 114, 115, 116, 117, 181
Diffusion flux XV, 44, 60
Diffusion model 46, 114, 121
Dimensionless coordinate 49, 51, 52, 54, 56, 57, 60, 62, 86, 102, 106, 110, 128
Direct integral method XVI, 152, 156
Dispersion XV, 15, 16, 17, 18, 30, 32, 33, 37, 47, 48, 50, 70, 82, 85, 86, 87, 108, 111, 207
Dispersion coefficient XV, 15, 17, 30, 37, 48, 50, 111
Divergence 76
Dynamic axial dispersion model 14
Dynamic viscosity XVI

Effective diffusion coefficient XV, 44, 76
Ejector 98
Ejector reactor 98
Elementary reaction 7, 153, 187
Emulsion phase 71, 72, 73, 74, 75
Endothermic reaction 46
Energy balance 9, 14, 20, 21, 23, 25, 26, 27, 28, 29, 37, 41, 46, 47, 51, 52, 55, 56, 57, 59, 61, 62, 63, 64, 73, 74, 92, 105, 106, 112, 113, 157, 176, 203, 207
Enthalpy 20, 21, 23, 27, 105, 176, 177
Enthalpy of formation 29
Enthalpy of reaction 22, 27, 29, 105, 146
Enzyme 8
Equilibrium constant V, XV, 7, 37, 135, 143, 146, 157, 180, 181
Equilibrium ratio 104, 117
Ergun equation 64
Euler method 34, 35, 75
Exothermic reaction 20, 46, 177
Explicit method 34, 213

Fanning equation 29
Fick's law 83, 103
Film theory 99, 100, 102, 114
Film thickness 100, 102, 208
Finite difference 65, 94, 95, 119, 120, 121, 211, 214
Finite difference method 65
Finite element 66, 119, 211
Finite element method 211
Fixed bed V, 41, 42, 45, 46, 47, 48, 49, 52, 53, 60, 62, 64, 65, 66, 67, 68, 69, 78, 79, 81, 93, 94, 128, 129, 132, 157, 158, 160, 178, 179, 180, 205
Flake 43
Fluid film 25, 41, 42, 44, 57
Fluidization 68, 69, 70, 71, 78
Fluidization velocity 78
Fluidized bed V, 41, 42, 68, 69, 70, 71, 72, 73, 74, 78, 79, 81, 82, 184
Flux XV, 27, 42, 60, 61, 63, 83, 84, 89, 90, 91, 100, 103, 105
Forward difference 212
Friction factor 29, 64
Fugacity 118

Gas–liquid equilibrium 117
Gas–liquid reactor 97, 98, 99, 101, 109, 112, 113, 118, 119, 121
Gas phase 76, 77, 78, 81, 87, 88, 92, 100, 108, 109, 110, 111, 112, 114, 118, 182
Gas-phase diffusion coefficient 76, 77, 113
Gas solubility 118
Generation rate XV, 6, 11, 22, 42, 63, 84, 90, 91, 124, 127, 129, 131
Gradient method 139
Gradientless reactor 130

Hatta number XVI, 121
Heat capacity XV, 21, 25, 26, 27, 28, 55, 56, 106, 176, 177, 179, 203
Heat conductivity XVI, 51, 55, 56, 59
Heat transfer coefficient V, XV, 28, 57, 59, 76, 77, 78, 79, 106, 135, 176, 179
Heat transfer parameter 37, 157
Henry's constant 118
Henry's law 118, 182
Heterogeneous catalysis 41
Heterogeneous catalyst 160

Heterogeneous model 46, 47, 52, 60, 61, 64, 66, 68
Heterogeneous reactor 9, 123
Homogeneous catalyst 9, 195
Homogeneous model 46, 47, 64
Homogeneous reactor 1, 9, 20, 28, 31, 32, 65, 92, 97, 153
Hot spot 68, 157
Hyperbolic partial differential equation 132

Ideal gas 14, 28, 29, 37, 49
Identifiability of parameter 2
Implicit method 34
Independent variable XVI, 132, 133, 135, 137, 144, 211
Initial guess 35, 139, 147, 211
Initial value problem 32, 34, 65, 66, 74, 87, 93, 94, 95, 211, 213
Intalox ring 99
Intalox saddle 99
Integral method 123, 125, 126, 152
Integral reactor 126
Internal energy 20, 21, 23, 26, 27, 55
Isothermal experiment 158
Isothermal reactor 14, 175, 211
Iteration 35, 137, 139, 148, 211, 212
IVP. *See* initial value problem

Jacobian 35, 137, 211, 213, 214, 216

Kinematic viscosity XVI
Kinetic constant 156
Kinetics V, VI, 1, 5, 6, 7, 8, 11, 17, 34, 36, 43, 58, 72, 91, 92, 98, 105, 106, 123, 126, 127, 128, 143, 148, 177, 189, 203, 205, 211, 218
Knudsen diffusion 44, 77, 181
Kunii-Levenspiel model XV, 71, 72, 73, 75

Laboratory experiment 123
Laboratory reactor 64, 99, 123, 131
Labyrinth factor. *See* tortuosity factor
Laguerre polynomial 65
Laminar flow 9, 29, 30, 31, 81
Laminar flow reactor 9
Langmuir–Hinshelwood kinetics 8
Le Bas increment 115
Legendre polynomial 65
Lessig ring 99

Levenberg-Marquardt method 139
Linear model 144, 145
Linear multistep method 34, 213, 215, 216
Linear regression 124, 126, 128, 144, 145, 147, 192
LSODE 216

Mass transfer coeffcient 74, 79, 102, 103, 104, 106, 113, 114
MATLAB 226
Mechanically agitated reactor 98
Michaelis–Menten kinetics 8
Michelsen method 214
Minimization 1, 45, 131, 135, 140, 150
Minimum fluidization velocity 78
Miniring 99
Molar flow 10, 12, 13, 16, 17, 18, 21, 24, 36, 48, 49, 50, 53, 60, 68, 110, 112, 128, 129, 130, 132, 135, 150, 175, 181
Molar volume XV, 23, 115, 116
Molecular diffusion 15, 30, 44, 77, 100
Moving bed 41
Multiphase reactor V
Multiple reactions 5

Newton-Raphson method 35, 91, 94, 137, 139, 211, 216
NLE. *See* non-linear equation
Non-elementary reaction 7
Non-isothermal reactor 157
Non-linear equation 34
Non-linear model V, 132
Non-linear regression 135, 156, 157, 204
Normalization of data 150
Numerical differentiation 63, 124, 126
Nusselt number XVI, 78

Objective function XV, 135, 137, 138, 139, 140, 141, 150, 151, 153
ODE. *See* ordinary differential equation
Optimization 1
Ordinary differential equation 10, 12, 31, 32, 64, 66, 71, 73, 74, 89, 93, 94, 132, 152, 211, 212, 215
Orthogonal collocation 65, 66, 94, 118, 121, 211
Orthogonal polynomial 65
Overall heat transfer coefficient 25, 52, 61, 176, 179

Packed bed. *See* fixed bed
Packed column 84, 99, 108
Pall ring 99
Parabolic partial differential equation 16, 52, 64
Parallel reaction 178
Parameter VI, XV, 15, 18, 35, 94, 100, 123, 126, 128, 132, 133, 134, 135, 136, 139, 140, 141, 142, 143, 145, 146, 147, 152, 156, 157, 158, 160, 176, 183, 185, 186, 190, 197, 203, 212, 218
Parameter estimation VI, 94, 126, 128, 133, 136, 140, 141, 143, 145, 146, 152, 156, 157, 158, 160, 190
Parameter sensitivity 141
Partial derivative 21, 35, 37, 137, 139, 211, 214
Partial differential equation 16, 18, 32, 93, 118, 132
PDE. *See* partial differential equation
Péclet number XVI, 17, 18, 19, 30, 31, 111
Penetration theory 100
Peng-Robinson equation 118
Perkins-Geankoplis equation 116
PFR. *See* plug flow reactor
Plate column 98
Plug flow 9, 14, 15, 16, 17, 23, 25, 31, 32, 33, 36, 47, 50, 51, 60, 68, 85, 86, 87, 88, 93, 94, 109, 111, 112, 128, 132, 175, 211
Plug flow reactor 25, 31, 33, 211
Polymerization 8, 41
Pore size 45
Porosity XVI, 43, 76, 78, 186
Power law 156
Prandtl number XVI, 78
Pressure drop 29, 37, 45, 52, 64, 68, 69, 70, 82
Production rate 178
Pseudo-homogeneous model 46, 47, 49, 64, 68, 91, 128, 178

Random pore model 76
Rate V, XV, 6, 7, 8, 31, 34, 37, 58, 91, 92, 97, 106, 124, 127, 128, 129, 133, 135, 141, 142, 143, 144, 145, 146, 147, 155, 156, 157, 175, 177, 178, 180, 181, 186, 187, 188, 189, 197, 200, 201, 203, 213
Reaction enthalpy XV, 22, 27, 28, 29, 105, 146, 176, 177

Reaction rate XV, 7, 13, 17, 20, 22, 37, 63, 127, 135, 153, 180, 182, 185, 186, 189, 195, 198, 206
Reactor design 183, 190
Reactor technology V, 82
Redlich-Kwong equation 118
Regression XV, 116, 124, 126, 128, 133, 134, 135, 139, 142, 143, 144, 145, 147, 149, 152, 192, 195, 197
Reynolds number XVI, 29, 31, 78
Rosenbrock-Wanner method 214
ROW. *See* Rosenbrock-Wanner method
RPM. *See* random pore model

Scheibel equation 115
Schmidt number XVI
Semi-batch reactor 11, 12, 13, 27, 98, 107, 132, 176, 185, 195, 200
Semi-implicit method 34, 37, 213, 214
Shape factor XV, 54, 55, 90
Sherwood number XVI, 78
Simultaneous reactions 151
Slab 43, 44, 55, 58
Slurry reactor 81, 91, 92, 93, 131, 151, 185, 200
Spherical particle 52
Sphericity 64
Spray column 98
Stability 34
Steady-state model 12, 47, 49, 62, 87, 88, 94, 95
Stefan-Maxwell equation 114
Stiff differential equation 66, 213, 216
Stiffness 34, 213
Stoichiometric coefficient XVI, 5, 22, 153
Stoichiometric matrix 6, 7, 151, 175
Stokes-Einstein equation 115
Surface renewal theory 100, 114

Tank reactor V, 9, 10, 11, 23, 27, 81, 82, 88, 89, 93, 94, 107, 113, 119, 121, 123, 127
Taylor series 214
Tellerette 99
Temperature gradient 20, 46, 47, 52, 63, 157
Thiele modulus XVI
Tortuosity XVI, 76, 186
Tube reactor 9, 23, 29, 46, 175

Tubular reactor 9, 10, 14, 15, 18, 25, 29, 30, 61, 175, 189, 211
Two-phase reactor 41, 76, 92
Tyn and Calus formula 116

UNIFAC 118
Uniform surface 71
UNIQUAC 118

Variance 30, 141
Velocity VI, XVI, 16, 18, 29, 30, 43, 64, 68, 70, 78, 206
Venturi scrubber 98

Viscosity 114, 115, 116, 117
Volume element 14, 15, 16, 20, 21, 22, 23, 42, 43, 47, 48, 56, 60, 61, 71, 72, 82, 83, 84, 85, 88, 90, 101, 102, 109, 110
Volume increment 116
Volumetric flow rate XV, 12, 13, 14, 17, 49, 127, 128, 177

Wakao-Kunii correlation 77
Wake phase 70, 71, 72, 74
Wetted wall reactor 98
Wilke-Chang equation 115, 116, 117

www.ingramcontent.com/pod-product-compliance
Lightning Source LLC
Chambersburg PA
CBHW082323220526
45470CB00008B/2388